TECHNOLOGY TRANSFER AND HUMAN VALUES

Concepts, Applications, Cases

Peter B. Heller

Manhattan College

UNIVERSITY
PRESS OF
AMERICA

LANHAM • NEW YORK • LONDON

ACKNOWLEDGMENTS

I am in the debt of several people who, in one way or another, have helped me greatly in this project. Among them I would like to mention Father George Cotter, M.M., and Sister Elizabeth Farley, R.S.C.J., who made their files available to me at the Mission Project Service. Also, Ton de Wilde and Robert B. Fricke of Appropriate Technology International, Edward P. Bullard of Technoserve, Dr. William N. Ellis of TRANET, Dan Santo Pietro of TAICH, and Janet Alecon of VITA for various forms of assistance. Too, Dr. John Wilcox of Manhattan College, and Joan McQuary of Columbia University Press.

I am especially grateful to Paul Bundick, consultant to the World Bank, for his comments on early versions of my work, and to the anonymous reviewer of University Press of America for constructive readings of the later versions.

Finally, I acknowledge the assistance I received in the form of a Tenured Faculty Development Grant at Manhattan College providing the seed money to engage in this project.

A note of thanks should additionally be made to Elizabeth McGehee Parella for her effort expended in typing this manuscript.

Naturally, I am solely responsible for errors of omission or commission.

P.B.H.

TABLE OF CONTENTS

Page

Part II. The Cases

PREFACE

Technology Transfer and Human Values is a specific focus of the broader Science, Technology, and Society programs that are finding increasing acceptance on college campuses and in the scientific community, judging by the proliferation of Technology and Values symposia and other evidence. It is also a subject of growing interest to organizations involved in the development process and even to businesses anxious to expand their reach in developing societies.

Briefly, the present work strives to help the understanding of sociocultural values with which all technologies are inevitably associated. It tries to do so especially with reference to appropriate and inappropriate technologies, suggesting that the range of social and cultural values incorporated into these technologies is far broader than the common criteria of economic growth or state-of-the-art technologies. Thus, freedom, quality of life, preservation of the environment, and the control of technical and social systems all find their place in the technology-values interface and highlight the role of "man" in the machine-man dyad. Hopefully, then, this work will throw light on the social role of technology.

INTRODUCTION

The purpose of this work is first to explain and
then to document by means of case studies from various
sources and covering a range of subjects the relationship between technology transfer on one hand and human
values on the other, a focus more specific than the
familiar science-and-society dyad to which it is
linked. The thrust of the work will suggest the extensive physical, economic, social, cultural, psychological, environmental, and political changes that
occur as a result of moving technology from one setting
to another, especially into a different human values
environment, as well as the impact which this context
will have on the technology itself.

Technology Transfer and Human Values:
A Symbiotic Relationship

This relationship follows from the premise that
society is not simply a product of its technology but
that its dominant economic, social, cultural, psychological, and political forces also guide the direction
of technological change. Technology underpins economic, sociocultural, and political systems while the
subsumed human values in turn shape the development
and application of technology since technical change
cannot be separated from the broader forces operating
in society. Accordingly, the interface between the
two is symbiotic.

The book demonstrates the extent of value choices
--indeed, value dilemmas and conflicts--underlying
technology transfer as technology selection is related
to an order of priorities in the physical, economic,
social, cultural, psychological, and political realms.
This suggests that technology-related choices are
value-laden rather than neutral and thus that there
are trade-offs that must be made between the benefits
of technology and other public policy goals subsumed
under the rubrics of international trade, development,
or modernization. More specifically, this work examines assumptions about standards of living, the quality of life, lifestyles, nature, the environment--in a
word, life--and, of course, technology.

Accordingly, both the mini examples woven into
the body of Part I and the full-length case studies in
Part II bring out how these multidimensional and interdisciplinary issues are resolved with reference to
diverse criteria as well as the political context in

which policy alternatives and priorities are determined.

Part I presents the various concepts of technology transfer and diffusion (Introduction and Chapter 1), the relevance of human values and culture in the process (Chapter 2), the various types of technologies transferred (Chapter 3), some common and not so common problems involved in the process (Chapter 4), public policy issues between different categories of countries (Chapter 5), and establishes a framework for the critical reading of the cases (Chapter 6).

Part II presents the full-length case studies, grouped into meaningful units to correspond with the chapter sequence. Such a breakdown blends the theoretical with the practical, presents cases involving both microtechnology and macrotechnology, in micro and macro settings, and at different levels of intensity and sophistication.

PART I

CHAPTER 1

TECHNOLOGY TRANSFER AND TECHNOLOGY DIFFUSION

Technology, Technology Transfer,
Technology Diffusion

Technology

The term technology has no all-embracing definition. It usually refers to the organization of all
scientific and empirical knowledge required for producing and distributing any goods or services. That
is, it is the direct application of science, invention,
industry, and commerce to matters deemed important to
a society's lifestyle. One of the many definitions in
the literature represents technology as "a probe that
reveals issues that might otherwise have escaped attention." Another holds that it is "the critical missing
catalyst to accelerated development." Still another
presents it as "knowledge about physical relationships
that permits some task to be accomplished, some service rendered, or some product produced."[1]

Technology includes production, organizational,
and managerial skills as well as what man does with
these--equipment, materials, installations, and systems. And indeed, most definitions highlight one or
more of these aspects of technology--its material or
nonmaterial form, its legal or systemic characteristics, its subject, method, type, or other factors.
None of these definitions are sufficiently indicative
of the extent to which technology is a catalyst or
force for change, not only in the way of doing things
and living standards but also in social institutions,
attitudes, expectations, and value systems.

Technology Transfer

Generally, technology is organized and replicable
knowledge. It also has a transmittable quality, that
is, technology can be transferred. Indeed, technology
can be transferred vertically, up or down, from one
institution, system, or stage of development to another. For instance, the U.S. Department of Agriculture can transfer a new technique down to grassroot
farmers or the Federal Bureau of Investigation can
transfer a law-enforcement method to local police
forces.

Technology can also be transferred horizontally,

1

laterally from a firm or sector or system to another--
say, from one multinational corporation's affiliate in
a country to another elsewhere,[2] or the U.S. National
Aeronautics and Space Administration (NASA) can and
has shared its findings in communications, integrated
circuits, gas turbines, cryogenic insulation, re-
chargeable cardiac pacemakers, solar cells, and so on
with other federal agencies.[3]

But technology transfer may also be _transnational_
(international), taking place between countries rather
than _intranational_ (domestic) occurring within the
same country or society. Too, technology transfer may
be _commercial_ (generally for profit) or _noncommercial_
(usually in the public area, for public use, whether
civilian or military). Furthermore, technology transfer may be _homophilous_, sometimes known as single
track, where the item transferred is used without modification. Or it may be _heterophilous_, or new track,
where an item is used in a novel activity, even one
for which it was not originally designed and therefore
has to be adapted not only to different conditions but
also different purposes--for instance, the diesel engine, originally used on submarines and only later on
railroads. Or, the transfer may involve _naked tech-
nology_--e.g., a semiconductor or device--as against
embodied technology--e.g., a complex instrument of
which the semiconductor is a key element.

Furthermore, technology transfer may be planned
or purposeful, as when a firm purchases a license to
use another's process. Or, it may be unplanned or in-
cidental, with little or no premeditation. The latter
is known as _technology diffusion_. When such an unin-
tentional process results in the adoption of a new
product or technique by the spread of ideas from its
creator to the ultimate adopter like "the ooze of
liquid seepage," the outcome may be similar to that of
technology transfer. In the literature, the two are
often used interchangeably.

But whatever the case, especially technology
transfer involves two actions and often two _agents_.
There is the one which sends and the one which receives. In many cases of transfer between unequal
societies (say, from developed to developing countries), the agent which sends (especially a multinational corporation, or MNC) is frequently the agent
which receives. The entire collection of private and
public organizations involved in technology transfer
is called the _technology delivery system_.

In the last analysis, however, there is no authoritative definition of technology transfer even though
some writers have been more specific than others.
Thus, while technology transfer has been defined as
"the generation and/or use of scientific or technological information in one context and its re-evaluation
and/or implementation in another,"[4] or as a "complex
and incompletely understood process [that] involves
communications in which the message contains technological elements,"[5] or more generally as the transformation of research and development into products,
technology transfer has a chameleon-like quality that
evolves to a degree as a function of participants,
applications, physical and sociocultural environments,
and problems.

Technology Transfer Process

Figure 1.1 below suggests that there are many
possible technology transfer agents. Domestically,
government agencies at different levels, technical in-
stitutes, private voluntary organizations and universities, and corporations are the most frequent transfer agents. Transnationally, because of the interna-
tionalization of their production, marketing, and distribution functions, multinational corporations are
the most common in the commercial (civilian) field.
The box in the diagram relates to the ways in which
technology can be transferred from supplier source to
recipient/user. Generally, the mode of transfer depends on the particular technology involved as well as
on the normal commercial considerations (technological
base, financial resources, licensing requirements)
associated with the transaction. Too, on the human
environment of the technology users.

Because of this, the efficiency of the transfer
method is heavily dependent on the level of communication and understanding, especially on a person-to-
person basis, between technology supplier and technology recipient. Indeed, the transfer of information
is a necessary precondition for effective results.
And the more complex the technology, the more necessary the communications and interpersonal relationships required to consummate the technology transaction. Accordingly, technology can best be transferred
by the movement of individuals rather than other means
of communications (patents, technical literature, re-
prints, etc.).

The transfer mechanism may be direct (the

3

FIGURE 1.1

The Technology Transfer Process

Transfer environment, viz.:

Economic conditions
Home and host government regulations

Source environment firm	Source Technology	Transfer agent such as multinational consulting engineers on-site training of foreign personnel technicians teachers students	User firm	Receiving environment

Source (technology supplier):

Private, for profit such as
commercial firms

Private, for nonprofit such as
research organizations
foundations
professional associations
academies and universities
labor unions

Public, for profit such as a
state export enterprise

Public, for nonprofit such as
a state research organization
an international organization
agency

User (technology recipient) such
as multinational
affiliate
other private firms
public enterprise
international organization
agency

Transfer Mechanisms

Turnkey plants (direct transfer)
Products (naked or embodied technology)
Licenses and patents (indirect transfer)
Emulation (reverse engineering)
Technological services, processes
Industrial espionage, other property violations
Publications
Meetings, seminars, colloquia, cooperative research
Education abroad or imported education at home
Site visits, on-the-job training
In-house transfers
Joint ventures

SOURCE: Adapted from Harvey W. Wallender III, Technology Transfer and Management in the Developing
Countries: Company Cases and Policy Analyses in Brazil, Kenya, Korea, Peru, and Tanzania.

4

establishment of production facilities in another location) or _indirect_ (the transfer of proprietary knowledge such as patented products or licensed processes). Or, it may be by such other modes as industrial and trade fairs, workshops and seminars, training programs at the technology supplier's location or that of the technology recipient. Each transfer mechanism is most appropriate to a particular purpose or environment.

Foreign direct investment involves full or partial ownership in a foreign subsidiary by the parent firm. The parent organization typically provides technology, capital, management, and marketing skills while the foreign subsidiary supplies material and labor resources. Foreign direct investments are effective channels for technology transfer because the equity interest in the foreign subsidiary by the parent firm is a powerful economic incentive to insure that the transfer proceeds efficiently.

However, from the standpoint of the recipient country, foreign direct investments are not very desirable because of the limited ability to learn and absorb the technology and management skills where the parent firm retains complete or majority ownership of the subsidiary. Hence the desire by recipient countries for shared equity or joint ventures.

A license agreement may be concluded between two parties that may or may not be affiliated. When technology is transferred through a licensing agreement between two affiliated parties, both foreign direct investment and licensing channels are used. Thus, foreign direct investment and licensing are not mutually exclusive processes.

In its broadest sense, a _joint venture_ can refer to any form of collaboration between two or more business entities. However, the term may also be used in a more specific sense to refer to the establishment by two partners of a third business entity for a specific purpose and whose equity and costs they share. Important reasons for this kind of technology transfer include accommodation to foreign investment laws (several governments prohibit wholly-owned subsidiaries in order to force some equity participation by local firms); the sharing of costs or skills of a technical, production, marketing, and/or managerial nature; risk reduction against such contingencies as financial liability or loss or nationalization; as a prelude to acquisition or merger of the partners; and

5

for accounting or tax reasons. But joint ventures involve disadvantages, too, focusing on the relative independence of action of each partner. Accordingly, joint ventures must be assayed against the alternative licensing arrangement. History documents both relatively successful (the Anglo-French Concorde supersonic jet plane) and unsuccessful joint ventures (the French-German-Dutch Unidata computer planned to compete with IBM).

In the case of the sale of <u>turnkey plants</u> abroad --those ready to go on line--the parent firm retains no equity interest. Here plants are built and sold as a package to the foreign recipient, often including all the know-how and skills necessary for plant operations by indigenous workers. Countries where only a limited technical and industrial infrastructure exists prefer to buy turnkey plants because these afford a quick and efficient method for acquiring manufacturing capabilities, especially when accompanied with the necessary training and technical assistance.

Licensing is usually technology specific and most frequently includes, besides the technology, know-how, trade secrets, trademark, or sales and distribution rights for a particular product. Licensing conveys to the recipient, or licensee, the right to use the technology or product for a limited time and application in return for a lump-sum payment, royalties in some fixed ratio to the resultant sales, or both.

The sale of technology-intensive products represents the dominant mode of technology transfer which itself is a major element of international business. From the viewpoint of the recipient, this is not very good because the important technological content is generally difficult to extricate even with <u>reverse engineering</u>, a euphemism for plagiarizing technology through the use of various methods. In contrast, this is the preferred mode of the technology source since manufactured products represent the highest value-added and therefore the most profitable embodiment of its technological capacity. Whether foreign direct investment or its most common alternative, licensing, is involved, multinational corporations figure preeminently in such transfer processes because "the MNCs combine superior management techniques, better product or manufacturing technologies, worldwide research activities, centralized authority structures, large financial resources, and good communications systems to bring technological solutions found in one

6

geographical area to bear on a problem or opportunity perceived in another."[6]

Among the other means of transferring technology, the role of consulting firms may be highlighted. They provide the diagnostic, evaluational, and judgmental skills necessary to analyze problems, identify technology suppliers, and provide the needed management and marketing skills in all ventures, especially from developed to developing countries.

Stages of Technical Advance

Figure 1.2 illustrates the stages of technical advance in relation to technology transfer.[7] The first workable model of a new product or process is often termed an invention (Stage II) and its first commercial production or use an innovation (Stage III). The former term is more comprehensive since it includes activities whose purpose is to develop innovations as well as other activities such as basic research with different objectives. Strictly speaking, innovation refers to any idea, practice, or material artifact introduced for the first time and that is discontinuous with past practice. The terms innovation and invention--even discovery--are often used interchangeably.

At this critical point, that of the first economic application of technology, the transfer will achieve an economic value as an innovation when it finds a use, an effective demand. At adoption (Stage IV), various forms of resistance (economic, ethical, moral, social, cultural, psychological, and so on) will have been overcome and the diffusion process--the spread or dispersal or dissemination or generalization of the innovation or its multiple use--occurs. For instance, the turbojet engine developed for aircraft was subsequently used for fog dispersal at airports.[11]

Technology diffusion, then, is the multiplier effect, even though unplanned, of innovation. Diffusion involves primarily production and marketing activities--say, tooling up, incorporating new techniques into the manufacturing process, improving distribution or selling procedures. It is during the diffusion process that the use of a new product or process grows toward an equilibrium level of utilization so that diffusion can be viewed as a movement from an old to a new balance. It should be noted that diffusion may proceed at several institutional levels. Too, that firms diffusing technology may be pioneers or imitators.

7

FIGURE 1.2

The Stages of Technical Advance

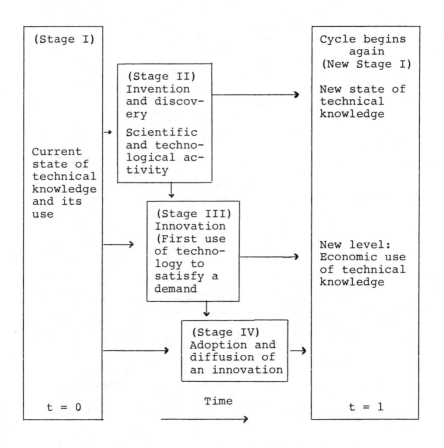

SOURCE: William H. Gruber and Donald G. Marquis (ed.),
          <u>Factors in the Transfer of Technology</u> (Cam-
          bridge, Mass.:   The M.I.T. Press, 1969),
          p. 257.

8

Pioneers are the first to use new technology successfully and thus the first to realize the savings in
production costs and the improvements in product performance made possible by the new technology. In contrast, imitating firms effect diffusion by duplicating
the successful production techniques exploited by
pioneering firms.

Both the innovative and diffusion processes are
essential for economic progress and more generally development. A device or process, no matter how great
the technological advance that it embodies, makes no
contribution until it is produced, marketed, and in
the hands of those who can use it. While technology
transfer tends to shorten the innovation time, the
speed and extent with which an innovation diffuses,
and in turn its impact on the industry and economy in
general, depend greatly on the magnitude of the technological advance. Consequently, it is not possible
to divorce the innovative from the diffusion process.
And in practice, the processes of technology transfer
and utilization are continuous, with all stages of
transfer activity taking place simultaneously.[8]

However, separate technological innovations tend
to occur in clusters and sporadically, and their major
effects are incremental. And when transfers are between different industries, they are inclined to be
sluggish because technology is often industry-specific.
But whatever the case, the cycles related to technology activities seem to be positively correlated with
business cycle fluctuations.[9]

Figure 1.2 does not indicate that for technology
transfer to be effective, for integrated technology
transfer to take place, know-how (learning to do) must
be accompanied by knowledge transfer (learning to
know). Otherwise, such superficial technology transfer as a mere change in the geographic location of a
plant will doom its recipients to continued dependence
on the technology supplier and thus to potential exploitation without yielding economic well-being and
social development.[10]

Technology Transfer Models

Figure 1.3, showing four models of technology
transfer, conceived by two engineers in private industry,[11] should help to conceptualize the process.
One final term appearing in Figures 1.2 and 1.3 that
calls for comment is technology adoption, because it

is related to the central concept of this work. Technology adoption usually occurs before technology absorption, which is successful utilization. As the sections below should make clear, in order to be successful the technology chosen must "fit" into its physical, economic, social, cultural, and psychological environment. The adoption of the technology implies that, at least in the perception of the technology user (assuming his sensitivity to this critical issue), the technology transferred can become part of the new setting and of the dynamics which characterize the operations of the systems of that setting. The transferred technology has to mesh with those systems and become an element of the technological, economic, social, cultural, and psychological fabric. If this does not happen, there is ultimately no transfer or no successful transfer, that is, long-term beneficial effect for its recipient.

Differences in countries' adoptive and absorptive abilities often explain why the technology transfer process is effective in some cases but not in others. Perhaps the most obvious example of the former involves Japan. Its success in using foreign technology has been attributed to its extensive technological infrastructure including skilled managers and engineers, a highly-trained work force, capital-intensive industries, and especially extensive research and development--the most important prerequisite for absorbing alien technology. Most developing countries lack these prerequisites, which explains why these technologically backward societies are least able to absorb imported technology effectively and to maximize the benefits of technology transfer.

For even if the technology recipient learns how to set up, operate, maintain, and repair a given piece of equipment and in general all tasks related to its handling, successful absorption would be far from assured. Indeed, to have this happen the technology user would have to develop stable relations with all sectors of the environment (suppliers of materials and energy, financiers, the marketplace, various government agencies), know what competitors are doing and what new developments are occurring, and so on. All these relations involve reciprocity. Thus, a technology is fully adopted only when the user has established stable relationships with all the elements of society necessary for support and when the balance among these relationships is favorable.[12]

FIGURE 1.3

BRIDGING AGENCIES

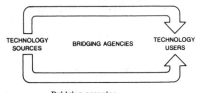

Bridging agencies.

R&D DIFFUSION MODEL

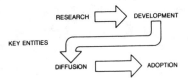

Research and development diffusion model.

PROBLEM-SOLVER MODEL

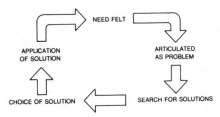

Problem-solver model.

TECHNOLOGY TRANSFER SUMMARY MODEL

Technology transfer summary model.

SOURCE: Louis N. Mogavero and Robert S. Shane, <u>What Every Engineer Should Know About Technology Transfer and Innovation</u> (New York: Marcel Dekker, Inc., 1982), pp. 2-3.

Nature of Innovation

Not indicated in the diagram is the fact that every innovation is a combination of ideas whose mental content is socially defined and whose substance is largely dictated by tradition. Still, the manner of treating the content, of grasping it, of altering it, and reordering it is inevitably a function of the individual mind. Thus, despite the differences in complexity between primitive and sophisticated cultures (see Chapter 2), there is abundant evidence of innovative ability in tribal societies. Even the African Bushmen, Australian aborigines, and arctic Eskimos have evidenced great ingenuity. Classical anthropologists point to the fact that primitive societies such as those just mentioned direct their innovations along different lines because the size and complexity of a society's cultural inventory establishes different limits, enabling these societies to, say, discover improved trapping devices while advanced societies discover photoelectric cells, computers, and nuclear power.[13]

Accordingly, what rates as intellectual superiority revealed in innovation is not simply a matter of biological inheritance since, throughout history, intellectual and/or scientific leadership have shifted among societies regardless of their genetic, ethnic, or national makeup. For instance, what some call "The American Century" referring to this country's preeminence in science and technology (as measured by the large number of Nobel Prizes won by Americans in these fields, for instance) is only a post-World War II phenomenon.

Historical Background

Technology transfer has taken place since the dawn of civilization. Sometimes technology transfer is slow. As mentioned, it took American railroads many years to adopt diesel power.[14] It has also taken the Swiss watch industry a long time to adopt quartz movements instead of those activated by mainsprings. But at times it is speedy. Consider how rapidly the electronics industry converted to solid-state technology from the use of vacuum tubes. Lately, as technology transfer has become a prism for a constellation of issues related to economic, technical, and industrial development in the Third World (that is, the less developed, or developing, or underdeveloped, countries) and international trade more generally.

12

interest in the process has grown accordingly and the tempo of transfer, helped by the communications revolution, has tended to accelerate.

Indeed, when technology transfer came to be viewed as at least a partial solution to the development problems of the economically underprivileged countries, it became fashionable and worthy of scrutiny by a variety of academic disciplines--economics, management, sociology, anthropology, history of science and technology, and political science (especially international relations and organization)--it being recognized that since the issues are multidimensional their treatment must be interdisciplinary. This heightened interest has spotlighted such issues as _dependencia_ (dependency),[15] imperialism, production factor substitution (especially as between labor and capital), the cost of technology and balance of payments effects, unemployment effects, barriers to technology transfer in the recipient societies, and in fact whether technology transfer is in the last analysis beneficial or detrimental to developing societies.

Westerners tend to take their own role of transferers of technology for granted. And in truth, the technology of sulphanamides, penicillin, DDT, transistors, television, computers, and the like was transferred from the West elsewhere through pharmaceutical companies, the U.S. Army, the World Health Organization, technical assistance missions, electronics firms, and national governments. Still, in earlier times, much of the technology used to flow from the East (Far East or Near East) to the West. For instance, a list drawn up to show which technologies were transferred from China to Western Europe across the centuries includes such items as the stirrup, the mechanical clock, the navigational compass, the caravel, and smallpox vaccine. The principal transfer agents were the Arabs, Turks, Portuguese, Crusaders, and other merchants and travelers.[16]

Even today, many processes are native to developing countries themselves and are available for transfer, on the basis of various compensation arrangements, to other developing or even developed countries. For instance, the Central American Research Institute for Industry in Guatemala has developed processing techniques involving several tropical fruits for the production of fruit juices, concentrates, or fruit puree preparations; or, the Central Food Technological Research Institute of India has developed a process

for the manufacture of baker's yeast from molasses.[17]
The National Water Department of Malawi has designed
a simple water pump, the "Maldev" (for "Malawi Development"), which can provide seven gallons of water daily
to 250 users, is easy to dismantle and repair, and
costs only $129 each compared to the much more complex
imported pump which is about $1,400.[18]

On a larger scale, industrializing countries like
Egypt are exporting technology. Egypt is doing so
mostly in the form of engineering, consulting, and
management services, technical assistance, and training,[19] while it, in turn, has acquired foreign technology.[20]

Albeit, since the developing world with 75 percent of the world's population but only 20 percent of
its income controls no more than 5 percent of aggregate scientific and technical potential,[21] the direction of the major flow of existing technology transfer
is self-evident.

Technology Transfer, Economic Growth,
and Development

Nobel Prize-winning economist Simon Kuznets characterized the relationship between technological pro-
gress--home-grown or transferred--and economic growth
as follows:

The major capital stock of an industrially
advanced nation is not its physical equipment; it is the body of knowledge amassed
from tested findings of empirical science
and the capacity and training of its population to use this knowledge effectively.
One can easily envisage a situation in which
technological progress permits output at a
high rate without any additions to the stock
of capital goods.[22]

This view, shared by other well-known economists
like Joseph Schumpeter and A. H. Hansen, means that
economic growth is now recognized to a considerable
extent as being the end-product of technological
change and transfer, that land, labor, and capital are
not its sole determinants as had long been the general
economic wisdom. One of these economists opined that
"technological knowledge possessed by people played a
bigger role in Germany's post-World War II recovery
than did the devastated and rundown plant which

14

survived the war."[23] For such technical progress--
which can be transferred--changes the production function and increases output through the introduction of
new products, new production methods, new markets, new
raw materials, and changes in the organization of an
industry.

Industrialization, highly dependent on technology,
is part of a unified, self-reliant development strategy inasmuch as many of the scenarios written for developing countries call for an industrial base to enable those societies to rise above their poverty and,
beyond economic growth, achieve such goals as diversi-
fied employment opportunities, greater economic choice,
improved infrastructure, and more equitable income
distribution--in short, development, defined sometimes
as a multidimensional process of social change.

Now, since development (in contrast to economic
growth) is not only about index numbers relating to
per capita or national income or about savings ratios
or capital coefficients but rather about and for
people and indeed about human values, it must begin by
identifying such human needs as adequate water, nutrition, schools, health care, secure livelihoods, and
the availability of suitable economic, social, political, and other institutions. To the extent that industrial growth is necessary to achieve these broader
developmental goals, technology transfer becomes crucial because it offers the recipients the opportunity
to realize greater returns without investing in research and development themselves.[24] Thus, failing
indigenous technology, technology transfer can substitute for or foster domestic innovation, underlie technological change, and in turn development.

But, as a country becomes more technologically
sophisticated--acquires more technological knowledge
and mastery,[25] makes greater technological effort--
about its own needs in technology transfer, it is
better able to achieve technology independence and
sustained development. Some technology recipients
(say, Japan, China, and the Soviet Union) have controlled foreign technology transfer strictly and have
used it to build and improve their own technological
base to replace imported technology. But others have
relied on the latter instead of promoting their own,
thereby also perpetuating their technological dependence on the suppliers of equipment, products, spare
parts, skills, and especially research and development.

15

Hardware and Software Technology Transfer

To the extent that the two can be distinguished, hardware technology refers to its physical embodiment such as infrastructure, plant, equipment, tools, and products, while software technology designates its nonmaterial elements such as knowledge of a particular design, process, or technique; know-how (the skill or ability to convert a particular technology into an economic reality); institutional structures; management; education, experience; legal provisions; financial incentives. In the transfer process the two types of technology are to some extent interchangeable in that additional or improved hardware can sometimes be substituted by the more efficient use, through better management or organization, of existing hardware. For instance, the bus system in New Delhi, India, was made more efficient merely through reorganization, without adding new equipment, and in the same country a fertilizer problem was solved in Punjab State by merging two agencies resulting in increased grain yields.[26] And indeed, development often calls for modifications or improvements in software rather than dramatic infusions of hardware.[27]

Yet, frequent preference for hardware is explained by the fact that it is easier to transfer equipment, machines, implements, tools, or devices from one society to another than nonmaterial items because technical and managerial procedures, legal and other organizational forms, and social values which, it will be seen, accompany technology are much more culture-specific and thus more difficult to move to a new setting. Too, software lacks visibility and is therefore less dramatic than hardware.

While examples of hardware technology transfers abound, some uncommon examples of software technology transfers can be found in the case studies developed by Accion International/AITEC.[28] These case studies indicate how development assistance to various Latin American countries can be effected through the transfer of a credit delivery system and management training--replicating models successful elsewhere in Latin America and lately in Somerset County, Maine, and improving the lot of the recipients. This was done by demonstrating how small loans (averaging $1,000 each, though hawkers and vendors receive as little as $300 each) and "management" training (involving primarily the keeping of records and basic marketing, planning, and other techniques) given to these "microbusinessmen"

16

can create jobs, increase earnings and profits, and lead to the repayment of the near-totality of the loans for further recycling. Since production usually includes both hardware and software, examples involving the transfer of both these types of technology are more commonplace. For instance:

> In the case of "Clafer Manufacturing" which produces stuffed toy animals in Colombia, a mechanic helped design a series of molds and patterns which reduces the time needed to turn out one stuffed animal from 15 days to some two hours. Later, the owner's husband built a special piece of equipment from old automobile parts powered by a washing machine that stamps out the textile pieces automatically and wastes less cloth in addition. Thus, instead of each worker laboring all day to trace and cut 90 sets of pieces for sewing into toy animals, one operator can now produce 300 sets a day, simultaneously eliminating the most boring, time-consuming, and costly part of the manufacturing process.[29]

It will be noted that this example illustrates more than one concept: The application of the mechanic's skills in a new setting (technology transfer), the use of a washing machine for a new purpose (technology diffusion), and the novel application of old automobile parts (innovation). But remember that the demarcation line among these terms is not always sharp or distinctive.

Another example of hardware and software technology whose transfer to developing countries would have significant implications is the following:

> In the use of microscopic techniques in the cattle-breeding industry, a new technique has been developed where calf embryos of microscopic size are split and combined, frozen in liquid nitrogen, and, months or even generations later, implanted in "surrogate-mother cows." This means that using genetic engineering methods, it is now possible to take two, three, even four different breeds of cattle chosen for various specific characteristics, join them together under a microscope, and in a single breeding cycle of just over nine months turn out an

17

animal that is a composite of them all.
Hence, thanks to this improved method of
artificial insemination and bio-engineer-
ing, it may be possible to transfer the
disease resistance of one breed to another
through this chimeric process, as it is
known, besides increasing the production
in milk and beef. In short, production of
superior offspring can be enhanced by
splitting an embryo surgically, implanting
it into a surrogate-mother cow, and thus
allowing the transplanted embryos to in-
herit from the surrogate-mother its required resistance to disease and other desirable characteristics.[30]

The implications of this technique to developing countries largely dependent on animal husbandry for income
are far-reaching. However, the technical difficulties
involved in the process may prove to be a restraining
factor in its transfer or its diffusion.

Reverse Technology Transfer

The term reverse technology transfer refers principally to the flow of technology from developing to
developed countries--especially to the movement of
trained personnel to more advanced societies--the so-
called brain drain. But the term can also refer to
the capital drain--the transfer of capital goods,
machinery, equipment, technical and managerial services from South to North, as these societies are
sometimes known collectively. An interesting study in
this respect involves Indian joint ventures abroad,
some of them in countries like Canada and other developed societies.[31]

What is significant from the viewpoint of development is that such reverse technology can be more
valuable, at least in the short run, than the flow
from developed to developing countries.[32] Indeed,
the emigration of scientists and technical personnel
from developing countries has occurred at the very
time when these feel compelled to seek experts from
developed countries.

The United Nations Conference on Trade and Development (UNCTAD) has published a series of case
studies in reverse technology transfer focusing primarily on the brain drain of professionals from those
countries where they have been heaviest, namely,

18

Pakistan, Sri Lanka, and India.[33] In nearly every
case the supplier country loses a valuable resource
(highly trained professionals) whose education is paid
for or subsidized by the home government while the
country of destination gains "ready-made" talent and
skills at little or no cost. It is widely known that
many physicians, nurses, technicians, engineers,
scientists, and other professionals from Third World
countries are found in large American, Canadian, British, and Australian cities among others.

In fact, it has been pointed out that even when
multinational corporations recruit local talent, there
is an internal brain drain that occurs as the latter
is siphoned off from home industries to the detriment
of the developing host countries.[34]

Reasons for the outflow of native expertise from
the less developed countries include the lack of social and political structures at home making possible
a reasonable stimulation of scientific activity. Also,
a distrust in local career prospects given the shortage of facilities, equipment, and scientific leadership or adequate national science policies. There is,
too, the attraction of higher earnings potentials
abroad, in developed but also some of the other developing countries such as the oil-rich states. Finally, there are all sorts of personal reasons.

By way of countermeasure, developing countries
are striving to improve the career and professional
prospects of such personnel as an inducement for them
to remain in their countries of origin, perhaps even
return home after migrating to the West where they enjoy not only much better compensation but especially
more advanced research facilities. For instance,
there has been a discussion of "delinking" professional education in developing countries from parallel
systems in developed countries which enable these
Third World professionals to find a ready job market
there. The proposal would make the educational system
in developing countries more adapted to domestic needs
--for instance, training paramedical personnel, even
midwives, to work in an integrated health service
dealing with disease patterns that are the product of
malnutrition, poor sanitation, and poverty,[35] rather
than highly skilled specialists whose expertise is
much more appropriate to the context of developed
societies.

At one point Communist Rumania announced

19

it would make all would-be emigrants under
retirement age repay the state in hard currency for the cost of free education they
received beyond the compulsory 10 years.
While officials were not sure why the regime of President Nicholas Ceausescu wished
to impose the emigration tax, it stands to
reason that, besides obtaining more badly
needed foreign currency, this measure would
have stemmed the flow of well-educated
Rumanians abroad, thereby benefiting host
countries with skills acquired at Rumanian
government expense. Under American pressure, the measure was rescinded in May 1983.[36]

FOOTNOTES

1. Respectively, Charles Weiss, Jr., Mobilizing Tech-
nology for Developing Countries (World Bank Reprint
Series No. 95) (Washington, D.C.: The World Bank,
1979), p. 1; Howard Pack, "Preface," The Annals of the
American Academy of Political and Social Science
(Philadelphia), vol. 458 (November 1981), p. 7; U.S.
Tariff Commission, Implications of Multinational Firms
for World Trade and Investment and for U.S. Trade and
Labor: Report to the Committee on Finance of the U.S.
Senate (Washington, D.C.: U.S. Government Printing
Office, 1973), p. 552.

2. David Benor and J. Q. Harrison, "Agricultural Extension: The Training and Visit Systems" (Washington,
D.C.: The World Bank, May 1977) (mimeo); also, Nawal
el-Messiri Nadim, Rural Health Care in Egypt (Monograph IDRC-TS15e) (Ottawa, Canada: International Development Research Centre, 1980).

3. Louis N. Mogavero and Robert S. Shane, What Every
Engineer Should Know About Technology Transfer and Innovation (New York: Marcel Dekker, 1982), pp. 2, 59-
62.

4. Samuel N. Bar-Zakey, quoted in Ralph I. Cole and
Sherman Gee, eds., Proceedings of the Colloquium on
Technology Transfer, 5-7 September 1973 (Silver
Spring, Md.: Publications Division of the Naval Ord-
nance Laboratory, 1973), Appendix, p. 80.

5. Bar-Zakey, quoting the Science Policy Research
Division, Congressional Research Service, Library of
Congress, Science Policy: A Working Glossary

(Washington, D.C.: U.S. Government Printing Office,
April 1972), p. 57.

6. U.S. Tariff Commission, Implications of Multinational Firms for World Trade and Investment and for
U.S. Trade and Labor (Washington, D.C.: U.S. Government Printing Office, 1973), p. 554.

7. William H. Gruber and Donald G. Marquis, "Research
on the Human Factor in the Transfer of Technology," in
Gruber and Marquis, eds., Factors in the Transfer of
Technology (Cambridge, Mass.: The MIT Press, 1969),
p. 257.

8. For a detailed study, see John E. Tilton, International Diffusion of Technology: The Case of Semiconductors (Washington, D.C.: The Brookings Institution, 1971).

9. Devendra Sahal, Patterns of Technological Innovation (Reading, Mass.: Addison-Wesley, 1981), pp. 228-
231.

10. Jacques Richardson, ed., Integrated Technology
Transfer (Mt. Airy, Md.: Lomond Publications, 1979),
p. viii.

11. Mogavero and Shane, What Every Engineer Should
Know About Technology Transfer, pp. 2-3.

12. Harold E. Hoelscher, Technology and Man's Changing World: Some Thoughts on Understanding the Interaction of Technology and Society (Beirut: American
University of Beirut, 1980), p. 36.

13. H. G. Barnett, Innovation: The Basis of Cultural
Change (New York: McGraw-Hill Book Co., 1953), pp. 16
et seq.

14. E. F. Hogan, "A Historical Study of Technological
Change in Railroad Motive Power" (Masters thesis in
industrial management, Massachusetts Institute of
Technology, 1958), quoted in Gruber and Marquis, Factors in the Transfer of Technology, p. 258.

15. The general dependency theory posits that because
technostructures--universities, research institutions,
laboratories, etc.--developed in the industrialized
world along with their productive capabilities and not
in the developing, previously colonized, territories,
the latter have come to depend on foreign technology

21

suppliers while failing to create their own local scientific and technical infrastructures to meet their needs.

16. Nicolas Jequier, Technology Transfer and Appropriate Technology (Singapore: European Institute of Business Administration [INSEAD], Seminar on the Management and Transfer of Technology, 1978), p. 7.

17. United Nations Industrial Development Organization, Technologies from Developing Countries (Development and Transfer of Technology Series No. 7) (Vienna, Austria: UNIDO, 1978), pp. 4, 8.

18. Robert Weller, Associated Press release, July 27, 1983.

19. Tagi Sagafi-nejad, Transfer of Technology from Egypt (document UNIDO/IS.362) (Vienna, Austria: UNIDO, 1982), p. 25. It is interesting to note that in the 30 cases of technology transfer, mostly to other Arab countries, Egypt's advantage is tabulated as being in political, commercial, or cultural links (14 cases), the cost of technology (7 cases), the quality of production (4 cases), and other reasons (5 cases)--ibid., p. 32.

20. For instance, steel-making from the Soviets-- United Nations Conference on Trade and Development, Transfer and Development of Technology in Egypt (document UNCTAD/TT/AS/7) (Geneva, Switzerland: UNCTAD, 1980), p. 24.

21. Surendra J. Patel, "Postface," in Richardson, Integrated Technology Transfer, p. 151. Patel is chief of the Transfer of Technology Division of the United Nations Conference on Trade and Development (UNCTAD).

22. Simon Kuznets, Toward a Theory of Economic Growth (W. W. Norton and Co., 1968), pp. 34-35.

23. Jacob Schmookler, "Technological Change and Economic Theory," The American Economic Review: Papers and Proceedings (Menasha, Wis.), vol. 55, no. 2 (May 1965), p. 333.

24. Paul Streeten, "Self-Reliant Industrialization," in Charles K. Wilber, ed., The Political Economy of Development and Underdevelopment, 2d ed. (New York: Random House, 1979), pp. 281 et seq.

25. This term is defined by Carl J. Dahlman and Larry E. Westphal as the ability to make effective use of technological knowledge in "The Meaning of Technological Mastery in Relation to Technology Transfer," The Annals of the American Academy of Political and Social Science (Philadelphia), vol. 458 (November 1981), p. 12.

26. Nicolas Jequier, ed., Appropriate Technology: Problems and Promises (Paris: Organization for Economic Cooperation and Development, 1976), p. 23.

27. For case studies in various training programs by Western expatriates in Africa, see Silvere Seurat, Technology Transfer: A Realistic Approach (Houston, Tex.: Gulf, 1979).

28. Paul Nevin, The Survival Economy: Micro-Enterprises in Latin America (Cambridge, Mass.: Accion International/AITEC, 1982). Accion International/AITEC is an independent, nonprofit organization dedicated to creating new opportunities for employment and economic well-being among the low-income populations of America. On a similar subject, see Marshall Bear and Michael Tiller, Microenterprise in the Outer Islands of Fiji (Washington, D.C.: Appropriate Technology International, 1982).

29. Nevin, The Survival Economy, pp. 9-13.

30. Wayne King, "Microscopic Techniques Have a Gigantic Effect on Cattle Breeding Industry," The New York Times, December 6, 1982, p. A6.

31. Ram Gopal Agrawal, "Third-World Joint Ventures: Indian Experience," in Krishna Kumar and Maxwell G. McLeod, eds., Multinationals from Developing Countries (Lexington, Mass.: D.C. Heath, 1981), pp. 127-130.

32. See, for instance, United Nations Conference on Trade and Development. The Reverse Transfer of Technology: Economic Effects of the Outflow of Trained Personnel from Developing Countries (document TD/B/AC,11/ 25 Rev. 1) (Geneva: UNCTAD, 1975).

33. Pakistan: Document TD/B/C.6/AC.4/3 of December 20, 1977; Sri Lanka: Document TD/B/C.6/AC.4/4 of December 19, 1977; India: Document TD/B/C.6/AC.4/6 of December 13, 1977.

34. Dmitri Germidis, Multinational Firms and

Vocational Training in Developing Countries (document SHC.76/CONF.635/COL.7) (Paris: United Nations Educational, Scientific, and Cultural Organization, 1976), p. 9.

35. El-Messiri Nadim, Rural Health Care in Egypt.

36. Bernard Gwertzman, "Rumania Plans to End Tax on Emigrants," The New York Times, May 19, 1983, p. A3.

CHAPTER 2

HUMAN VALUES

A value is an enduring belief that a specific mode of conduct or end-state of existence is personally or socially preferable to an opposite or converse mode of conduct or end-state of existence. Human values are those that derive from society and its institutions and personality.[1] They are images, representations, or real objects which can be perceived by individuals as habitually worthy of desire, volition, emotion, even passion, and which may move the individual to act. Human values are rooted in and in turn color people's perception of reality, of what is. Human values, often reflecting a given ideology, can be conservative, progressive, racist, sexist, or of any other coloration. At any given time some of these values can be in conflict with each other--e.g., Western generosity but also thrift; Christian versus business ethics; German militarism versus romanticism in the past.

Culture and Society

Culture is a somewhat broader concept than human values because, in addition to these, it factors in more objective elements such as knowledge, language, literature, arts, and the like which refer to the expressive and symbolic dimensions of society. Culture, then, includes the totality of customs among a given people and an abstracted common denominator of artifacts, ideas, and behavior. It was once defined as that complex whole which includes knowledge, belief, art, morals, law, custom, and any other capabilities and habits acquired by man as a member of society. In fact, however, the meanings of culture and human values are imprecise and some overlap between them is obvious.

To some extent people share common values; but they also hold different values or possess the same values to different degrees. While there are disagreements about the ranking of any set of values and the needs on which they are based, generally such values are related to the satisfaction of various human needs, material and nonmaterial. To wit, material needs have to do with survival, health, welfare, employment, etc.; sociocultural needs with cultural autonomy, distributive justice, various types of freedom; psychological needs with identity, creativity,

25

self-actualization, togetherness, and other forms of personal fulfillment; environmental needs with resource conservation, ecological preservation, ecosystem integrity; and so on. By extension then, human values refer to the attitudes, preferences, normative goals, symbolic universe, belief systems, and meanings which people assign to life in a given society or culture. The consequences of human values are thus reflected in all sociocultural issues, including that involving the transfer of technology from one setting to another.

Westerners, today the major suppliers of technology, cherish certain basic values and have embedded these into their technology--say, rationality, efficiency (productivity), functionality, planned obsolescence, professionalism, a belief that most problems have a solution (especially through the use of science and technology held in highest esteem rather than civilization or justice), and their corresponding desire to control nature and human institutions. But also the "virus of acquisitiveness and the idolatry of material success" and the related consumption-oriented society which considers more to be better, endless growth to be both possible and desirable, and this year's "new and improved" product superior to last year's.

The threat that modern technology may pose to humanity was forcefully described by the eminent French sociologist, Jacques Ellul, who pointed out that technology has replaced nature as the context of societal perception and decisions, that "technique" is an organizing principle in all human affairs, that the processes of technological change are irreversible, but yet that life is not happy in a civilization dominated by technology. In his view, technological progress is gradual dehumanization--a busy, pointless, and ultimately suicidal submission to technology even in the Western world.[2]

Still, in recent decades the earlier Western faith in rationality and utilitarianism may have waned, and questions have been raised about technology's impact on mankind and its environment beyond the fears expressed by Ellul.[3] For instance, a management consultant has warned that the increasing technology in American society will have to be balanced by the greater fulfillment of the spiritual demands of human nature; that the high tech of the computer age presently revolutionizing production methods

26

and society will have to be laced with more togetherness, with "high touch"; and that appropriate technology will have to be accompanied by appropriate sociology, "networking," providing a form of communication
and interaction suitable for this energy-scarce, in-
formation-rich future.[4]

Such trends have already been in evidence. A
sample poll was taken among American college students
and they ranked values such as interpersonal understanding ahead of money, comfort, and economic well-
being.[5] This seemingly paradoxical nonmaterialist
streak in Western values is explained by the fact that
of the five commonly accepted hierarchy of human needs
--survival, security, belonging, self-esteem, and
self-actualization--three of them (social needs, ego
needs, fulfillment needs) relate to the nonphysical
universe in contrast to the other two (physiological
needs, safety needs).[6]

Still, the recipients of Western innovations--
many of them Third World countries with essentially
non-Western cultures--generally tend to revere different values, make different assumptions about their own
particular traditions, rituals, kinship patterns,
friendship, contemplation, recreation, and esthetics
among others. This is because human values and more
generally culture change only slowly, often impercep-
tibly, so that the value system plays an important
role in preserving a society. Thus, conservatism
serves as a brake on too rapid a change, usually slow-
ing the process to the point where the society can
assimilate innovations without threatening its basic
structure. For whenever a culture is faced with an
innovation its members will wonder, no matter how sub-
consciously, not so much whether the new product or
process will work as what effects it may have on their
belief system, on their human values. By inference,
then, resistance to change is built into many cultures,
especially traditional ones, and this tenacity of
value orientations is a point to which the planning
and execution of technology transfer should pay great
attention.

Non-Western societies are less scientific, less
rationalized in their approach to reality. This difference with Western societies was highlighted in a
study that examined the receptivity of schoolchildren
in Nepal to scientific education:

It was found that these children continued

to adhere to traditional, that is, myth-
or folk-oriented, ways of explaining
several natural phenomena even though on
being questioned they also gave the
school-oriented version based on scientific principles. For instance, despite
their courses in science, a number of the
children continued to say that the earth
sat on the back of a fish or an elephant
and that earthquakes occurred when these
shifted their weight. On further ques-
tioning, they allowed that earthquakes
could also be caused by pressures in the
earth's molten core contributing to
stresses in the crust. To these children,
even after exposure to Western education,
the source of knowledge was immutable and
was transmitted from generation to generation. Thus, nature could be controlled
through religious ritual or magic. Ergo,
culturally rooted perceptions are stub-
bornly resistant to change through ordinary indoctrination.[7]

It will be recalled from Figure 1.1 and the related
discussion in Chapter 1 that the introduction of foreign educational curricula represents software technology transfer. One may deduce from this example, however, that the science-based cause-and-effect relationship congenial to Western minds was not so in
Nepal's traditional setting even in an academic situation.

Technology and Values

A particular technology, then, chosen to have
certain objectives also has values built into it since
it is part of the supplier society's culture. Whether
the transfer agent or recipient are conscious of the
fact or not, that technology subsumes certain priori-
ties--technical, economic, social, cultural, psychological, and political. This is so because the kind
of technology we choose influences the kind of society
we have.[8] To use Goulet's terms, "there exists a
value universe which is proper to modern technology."[9]
Or, to quote E. F. Schumacher who, it will be seen,
has done seminal work in this field, "each particular
type of technology is itself a political and social
shaping agent, with far-reaching sociological consequences."[10]

28

Thus, the technology that is transferred is like genetic material in that it carries within itself the core of the society in which it was conceived and nurtured. When it is released into the new environment, it tends to reproduce that society unless there is some kind of resistance to this form of sociocultural intrusion. And by being value-laden, technology transfer has a differential impact on different groups of people and environments, benefiting some more and others less, but always differently. Thus, technology often encodes old value differences of race, class, gender, and age even when transferred into supposedly advanced systems.

For instance, the introduction of airplanes into Western, especially American, society has probably had the effect of substantially increasing the number of business-related trips, while the transfer of civil aviation to the Arab world, a traditional society, has raised the number of kinship-related visits, benefiting different classes of technology-consumers in each case.[11] This suggests once more the basic theme presented here: That technology is shaped by societal forces as it is by technical ones and that society is both a product of its technology and vice versa, the two interacting in symbiotic fashion.

It may be instructive to mention an anecdote which, following extrapolation, would reinforce the concept that technology is value-laden.

> A mathematician's eight-year-old daughter characterized prime numbers (those which cannot be divided by any other integer except one, such as 11, 19, 83, etc.) as "'unfair...because there's no way to share them out evenly.'"[12] Yet, numbers do not change people nor can they be changed by them. So, if numbers can be "unfair," how much more so technology which can do both!

And yet, while it is relatively easy to accept the concept that the transfer of some technology--say, the transplanting of a crop in a foreign land--needs an appropriate physical environment (suitable soil, climate, water regime, and so on) to be successful, that variations in factor endowments (say, the relative ratio of labor versus capital) make some transfers economically cost-beneficial and others not, the issue of the compatibility of social and cultural factors was at first overlooked in decisions about what to use or produce and how to do it even though this connection between technology and values had been discussed for years. Consider, for instance, this West German observer's impressions of Indian attitudes vis-a-vis Western technology on the basis of his experience on the site of the giant Rourkela steel plant built in India by German firms in the late 1950s and early 1960s:

It is interesting to note that many Indians are unaware that the technical and industrial superiority of these foreign countries is set against a background of corresponding intellectual and civilizing processes. They seem to think that they can isolate the technology, the machines, the experience of the industrialized countries from a general setting they regard as largely useless and import them as they stand. They want to take over the functioning of Western civilization ready-made, that is, its products and the processes by which they are produced--in a word its "know-how"....[13]

Accordingly, the success or desirability of technology transfer is often far from assured even though, because of the close association of technology transfer and developmental goals, Third World countries are often willing to emulate, even mimick,[14] the beliefs, world views, prejudices, behavioral patterns, or lifestyles of Western cultures at the expense of their own.

This often generates a crisis of self-image and identity through the artificial elimination or at least repression of sociocultural differences. The example of the Shah of Iran's "modernization" program comes to mind. And when it happens, the recipient society may have to put up with technology that is violative of the environment, inequitable in its distributive effects benefiting one group at the expense of another, depletive of natural resources, offensive to a traditional way of life, hostile to cultural identity or dignity, undemocratic, and nonparticipatory.

Goulet describes a preliminary plan by Arthur D. Little, Inc., international consultants based in the United States, to establish a cold chain system to store frozen foods in Brazil. But he criticizes

30

the fact that this system was premised, in
part, on single-family units being equipped
with necessary refrigerators and freezers,
thereby automatically eliminating from
among potential beneficiaries the underprivileged who suffer most from food spoilage but cannot afford these appliances.
Ergo, a bias in favor of meeting the wants
of those with present or future purchasing
power is thus implicit in the technological
diagnosis itself of the problem.[15]

This is because, as Goulet puts it, technology is not
only "the bearer but also the destroyer of values";
technological innovation brings new values but as it
does so it displaces old ones and creates a new set of
determinisms. Social bonds, traditions, acts become
subordinated to criteria of performance or power as
technology imposes its own rationale, its own rhythms.

The catch, however, is that for the proper functioning of a given technology, the compatibility of
the recipient society's environment with the transferred hardware or software is crucial. For a modern
plant to operate successfully in a Third World setting,
it is also necessary to transfer a new outlook on factories there. Perhaps an analogy can be struck in the
realm of political science. After securing their independence, a number of Third World countries tried to
import packaged Western political systems (software)
with such typical features as constitutions, loyal
opposition, checks and balances, free speech, systema-
tized alternation of power, elections, and so on, with
disastrous results because their political cultures
rejected such incompatible alien implants.

This is true to such an extent that many definitions of technology increasingly factor in a human
values dimension, as when it is held to be "the use of
scientific knowledge of a given society at a given
moment to resolve concrete problems facing its development, drawing mainly on the means at its disposal, in
accordance with its culture and scale of values."[16]
Accordingly, technology is more than industrial or
applied science in a routine sense. It is a socio-
technological phenomenon where material and artifact
changes are accompanied by structural, social, cultural, and psychological changes as well. And it is
precisely because technology is value-laden that it
demands cultural receptivity and that its transfer is
problematic and unpredictable.

31

Thus, at the macro level some societies reject the increasing use of technology for the sake of maximizing other values. For instance, post-World War II Britain focused on social change and income redistribution rather than industrial and economic growth as was true of Japan in the same period. At the micro level, a community such as a village may resist the introduction of a new water distribution network (hardware) or new medical technology (software) in defense of traditional ways of doing things or on account of such factors as superstition. And incidentally, it was suggested that the potential sociocultural consequences, often anticipated ahead of time, of macro-technology like the construction of the huge Kariba Dam on the Zambezi River (or even the introduction of oil technology into Saudi Arabia) are not necessarily greater than the sociocultural upheaval, frequently unplanned for or even not always assessed after the fact, resulting from microtechnical projects such as the introduction of flashlights, wrist watches, bicycles, and more recently snowmobiles.[17]

Technology Occasionally Seen As Neutral

Whereas blind belief in technological determinism such as that evidenced by the classical sociologist William F. Ogburn over 60 years ago who saw the "pre-eminence of material culture as a factor in changing society today"[18] has lost some of its edge, some still view technology as neutral, unrelated to any consideration outside the state of scientific or technical knowledge.[19] These observers claim that technology has a self-contained completeness about it rather than being socially or culturally based. Since in this value-neutral view technology does not incorporate such biases, the technology is not "good" or "bad" in itself. Any beneficial or harmful effects flowing from technology transfer are attributed to the motives of those making the selection or the goals to which the technology is applied. Accordingly, failures can be blamed on inadequate social policies or a lack of sophistication in controlling the effects of technology transfer. Yet, even the determinations of the "free" market reflect assumptions and priorities through the relative demand-supply intensities which prices indicate.

Adding to the strength of the technology-is-value-free argument is the fact that some technology recipients have indeed managed to extract the transferred technology from its economic-sociocultural matrix.

For instance, China and Japan have been successful at appropriating Western techniques, sometimes by combining borrowed with indigenous technology, without allowing their own structures and cultures to be undermined. And when the Soviet Union acquires American technology, it is able to launder out the production relations and values of U.S. capitalism--high union-bargained wages, the protection of private interests, profit-maximization, at times conspicuous consumption --even though, judging by the rate at which Pepsi Cola or Chanel No. 5 have caught on in Eastern Europe, one should allow for exceptions even here. Generally, however, such destructured technology must be restructured in the living context--the material and human environment, culture, and institutions--of the recipient societies.

That is why more would tend to agree with the opinion that

> ...the neutrality of science is no longer
> taken for granted....Science performed
> within a particular social order reflects
> the norms and ideology of that social order.
> Viewed thus, science ceases to be regarded
> as autonomous and is seen as part of an in-
> teracting system in which internalized ideo-
> logical assumptions help to determine and
> form the very experimental designs and
> theories of the scientists themselves.
>
> It reflects the economic base and
> power relations of the society which has
> given rise to it....[20]

Consider the role of human values in the following examples. The first four of these deal essentially with hardware while the last two with software:

> The plastic sheets in West Africa covering
> the traditional roof may be technically
> appropriate in a hot climate, but one of
> the alternatives--the tin roof--is not only
> technically simpler but socially more pres-
> tigious since in some developing countries
> it is one of the most conspicuous signs of
> affluence and a top priority in a family's
> investment.[21]

> The second involves the Quechua In-
> dian peasant communities in central Bolivia.

33

When one of their villages organized a
cooperative for the production of cera-
mics, it decided not to introduce small
electrically powered kilns even though
this would have improved the product because of the even temperature an electric
kiln can generate. But it was rejected
because the electric kiln alternative
would have involved bringing a portable
generator to the village over which very
few villagers would have had technological mastery. Rather, the cooperative decided to adopt a kerosene-fueled oven and
simultaneously to try and improve the refractory (heat-insulating) properties of
local clay. The latter technology was
preferred because (1) even the poorest
villagers could afford the kerosene oven
whereas only the cooperative could have
paid for the electric generator; (2) the
villagers had had prior experience with
kerosene ovens.[22] The most efficient
technology (from the viewpoint of quality
and productivity) was rejected in the interest of other, more important and compatible, values.

Or consider the effects of introducing the snowmobile to the nomadic Lapps
who live within the Arctic circle in northeastern Finland. This Western gadget has
changed the relationship between the herds-
man and his reindeer, possibly affecting
the size and health of the herds. For
snowmobile herding precludes the earlier
man-reindeer relationship when these Lapps
used to tame their animals. Too, the use
of the new fuel-driven vehicle has ex-
tracted its toll from the ecology (noise,
accidents, the decimation of some fauna
because of the ease of hunting with a
powered vehicle compared to skis or dog
sleds). Furthermore, the status of older,
experienced men has declined since the
hazards of the snowy wastelands have been
reduced; and a new social stratification
has emerged between those families which
can afford a snowmobile and those which
cannot.[23]

Ponder, too, the effect of Western-

34

style, high-rise housing designed for nuclear family units in Singapore where this technology has undermined the traditional Chinese extended-family lifestyle.

Or how the introduction of modern medicine into the West Sahel region during French rule, the excessive boring of wells for the growing human and animal population, and the consequent overgrazing of pastureland have greatly contributed to the ecological disaster of the Sahelian drought.[24]

Questions have even been raised about the benefits of the highly-touted green revolution and its use of high-yield variety (HYV) seeds. For while it has increased cereal productivity spectacularly in some countries, it has also brought with it fertilizer pollution, the growth of agrobusiness at the expense of small farmers, but especially a more lopsided income distribution because of the unequal access to land, fertilizers, water, and credit, reinforcing the economic and political power of landholders and the urban industrial class who collect most of the surplus resulting from the use of green revolution technology.[25]

These cases all demonstrate the importance of human values in the equation even though such significance manifests itself differently. In some of the cases mentioned, the sociocultural values in technology transfer happen to harmonize with what is technically desirable (the tin rather than plastic roofs, the kerosene- versus electric-fired generators) and sometimes obviously not (well-boring in the Sahel). In the other cases (the snowmobiles, high-rise housing, and even high-yield variety seeds), there are serious trade-offs to be considered between the technical and the sociocultural benefits to be derived from such technology transfer.

Now consider the case of the large American multinational corporation which developed a highly desirable, inexpensive, high-protein instant cereal for sale in Latin America. In one country, the innovation was a complete failure because it had deliberately been marketed as a poor people's food.

This mini-example raises the interesting question of why there has been more sociocultural resistance to some of the technology transferred rather than to some other. One study has tried to answer this question with reference to European transfers of technology to India in the 16th and 17th centuries. During that period, Western artillery and shipbuilding techniques were accepted by the Indians, clockware and horse-drawn carriages were not, and there was indifference to the European printing press. It seems that what underlay acceptance or rejection of the various techniques was perceived threat in the former case but the belief of superior native technology in the latter. Overall, the Indians' response to cultural innovations (e.g., Western attire) was slower than to technological innovations, except when there already existed an appropriate local alternative to the technology. Seemingly, in all cases the Indian response was scrupulously selective depending on convenience, utility, exigencies, or other material or pragmatic considerations rather than "plain xenophobia, racial aversion, ingrained conservatism, or the pejoratively referred to 'oriental' resistance to innovation."[26]

It must be recalled that this is an Indian perspective. The West German's observations of Indian attitudes today, it will be recalled, was quite different, since he stressed their extremely nationalistic attitude distorting their view and assessment of technology transfer.[27]

Finally, consider this generic case of reverse engineering (a form of technology transfer) reported in the press:

It seems that Taiwan, branded as the "counterfeiting capital of the world," has just revised its law to reduce trademark, copyright, and patent infringements of foreign company proprietary technology covering anything from pens, watches, clothing, chemical processes, tapes, books, fast-food chain names to automobile parts. In the case of computers, the commercial pirates had marketed their wares as follows: "The best way to describe this computer is to say that it's an Apple. We copied it

down to the finest detail. I mean we got
all the software you ever want,...."[28]

While it is clear that such technology transfer
through property violations is explained by the economy's insufficient research and development or marketing or investments or formerly inadequate laws, consider the diversity of responses that such practices
by pirates of the business world might meet in other
sociocultural settings. In other words, even reverse
engineering--plagiarism by any other name--needs a
propitious environment to thrive, or there will be no
technology transfer in this form.

Summation

This chapter has stressed why, besides its other
attributes, technology should be considered to be
value-related and consequently why technology transfer
cannot be successful without an awareness of the human
values subsumed in the technology involved and the
technology-recipient society.

To the extent that technology and values are thus
interacting--and it is the contention here that they
are--the process of adaptation following transfer
strives to bring the two to a new equilibrium, enhanc-
ing the likelihood of the transferred technology's
acceptability. The evidence seems to suggest that at
least in the longer run and despite some resistance,
technology recipients especially in developing countries have been "buying" not only Western hardware
and software but with them the consumer-oriented
values underlying the predominantly capitalistic economic systems turning out these products and processes. These economic systems stress advanced industrialization, a highly consumer-oriented society, the
profit motive, a "business culture," and even the
English language that goes with it and all of which is
diffused by the Western technology-rich mass media,
especially advertising.

Whether such an evolution, if it is occurring,
adds up to cultural imperialism undermining local
identities, values, and lifestyles and if so whether
such a trend is indeed undesirable are still being
debated in a general way. However, there is no question about it among those who belong to the appropriate technology movement (see Chapter 3).

FOOTNOTES

1. Milton Rokeach, The Nature of Human Values (New York: Free Press, 1973), pp. 3, 5.

2. Jacques Ellul, The Technological Society (translated from the French by John Wilkinson) (New York: Random House, 1964).

3. Cutcliffe, Stephen H., et al., eds. "Introduction," Technology and Values: A Guide to Information (Detroit, Mich.: Gale Research Co., 1980), p. xvi.

4. John Naisbitt, Megatrends: The New Directions Transforming Our Lives (New York: Warner Books, 1982), pp. 40, 193.

5. Norman C. Dalkey, et al., Studies in the Quality of Life (Lexington, Mass.: D.C. Heath, 1972), p. 71.

6. Ian G. Barbour, Technology, Environment, and Human Values (New York: Praeger Publishers, 1980), pp. 59-80; but especially Denis Goulet, The Uncertain Promise: Value Conflicts in Technology Transfer (New York: IDOC/North America, 1977), pp. 15-30 and 243-251.

7. Francis E. Dart and Panna L. Pradhan, "Cross-Cultural Teaching of Science," Science (Washington, D.C.), vol. 155, no. 3763 (February 10, 1967), pp. 649-656.

8. Isao Fujimoto, "The Values of Appropriate Technology and Visions for a Saner World," in Richard C. Dorf and Yvonne L. Hunter, eds., Appropriate Visions: Technology, the Environment and the Individual (San Francisco, Calif.: Boyd and Fraser, 1978), p. 170.

9. Goulet, The Uncertain Promise, p. 16.

10. Cited in Witold Rybczinski, Paper Heroes: A Review of Appropriate Technology (Garden City, N.Y.: Anchor Press/Doubleday, 1980), p. 21.

11. Janet Abu-Lughod, "Some Social Aspects of Technological Development," in Mujid S. Kazimi and John I. Makhoul, eds., Perspectives on Technological Development in the Arab World (Detroit, Mich.: Association of Arab-American University Graduates, 1977), p. 41.

12. Quoted in Horace Freeland Judson, The Search for

Solutions (New York: Holt, Rinehart and Winston, 1980), p. 2.

13. Jan Bodo Sperling, The Human Dimension of Technical Assistance: The German Experience at Rourkela, India (translated from German by Gerald Onn) (Ithaca, N.Y.: Cornell University Press, 1969), pp. 41-42.

14. Denis-Clair Lamberg, Le mimetisme technologique du Tiers-Monde (Paris: Editions Economica, 1979), p. 7.

15. Goulet, The Uncertain Promise, pp. 102-103.

16. Organization for Economic Cooperation and Development, North/South Technology Transfer: The Adjustments Ahead (Paris: OECD, 1981), p. 18, quoting OECD/Interfutures, "The Problems of Technology Transfer between Advanced and Developing Societies," Midway Through Interfutures (Paris: February 1978), ch. 12.

17. H. Russell Bernard and Perrti J. Pelto, eds., "Introduction," Technology and Social Change (New York: Macmillan, 1972), p. 5.

18. William F. Ogburn, Social Change, with Respect to Culture and Original Nature (New York: W. B. Huebsch, 1922. Rev. ed. New York: Viking Press, 1950).

19. Joan Lipscombe and Bill Williams, Are Science and Technology Neutral? (London: Butterworths, 1979), especially pp. 18-26.

20. Mike Cooley, Architect or Bee? The Human/Technology Relationship (Boston, Mass.: South End Press, 1980), p. 55.

21. Jequier, Technology Transfer and Appropriate Technology, pp. 95-96.

22. Goulet, The Uncertain Promise, pp. 119-120.

23. Rybczinski, Paper Heroes, pp. 158-159; also, Perrti J. Pelto and Ludger Muller-Wille, "Snowmobiles: Technological Revolution in the Arctic," in Bernard and Pelto, eds., Technology and Social Change, pp. 166-199.

24. Nicholas Wade, "Sahelian Drought: No Victory for Western Aid," Science (Washington, D.C.), vol. 185, no. 4147 (July 19, 1974), p. 237.

25. United Nations Conference on Trade and Development, Transfer of Technology: Its Implications for Development and Environment (document TD/B/C.6/22) (Geneva: UNCTAD, 1978), pp. 27-28.

26. Ahsan Jan Qaisar, The Indian Pesponse to European Technology and Culture (A.D. 1498-1707) (New Delhi, India: Oxford University Press, 1982), p. 139.

27. Sperling, The Human Dimension of Technical Assistance.

28. Steve Lohr, "Crackdown on Counterfeiting," The New York Times, May 7, 1984, p. D12.

CHAPTER 3

TYPES OF TECHNOLOGIES TRANSFERRED

Despite the economic growth experienced by developing countries in the last two decades, poverty, human misery, and inequality are still very much a reality in the Third World. This has not only given ammunition to dependencia theorists who point out how the earlier colonial exploitation is now continuing under different forms. These critics also observe a dependency sequence beginning with purchasers needing outside sellers for an array of goods and services, then for capital (loans or grants), then for hardware and software technology (especially managerial expertise), and finally for access to markets. This process delays or prevents self-reliance, especially in those countries which need "expensive" Western technology transfer to promote their "modernization." It also raises the question of the suitability of the technology transferred with reference to various conditions prevailing in the transferee's environment, including the recipient's value system.

Alternative Technologies

This is the generic term used to describe a whole array of technologies that are viable substitutes for mainstream, usually modern, technologies. The technologies described below--and others which are constantly sprouting up to meet new needs--are specific instances of alternative technologies. As with several other terms presented so far, these, too, do not have precise parameters. Many are used loosely and often interchangeably, minimally overlapping with each other.

Intermediate Technology

The term intermediate technology (or IT) was first used by Dr. E. F. ("Fritz") Schumacher in his seminal book, Small Is Beautiful.[1] The German-born British economist and founder of the London-based Intermediate Technology Development Group (ITDG), who had been inspired by "Buddhist economics" and Gandhi's principles of respect for all life, the need for compassionate acts, service, shared labor, and simple industrial tools, turned intermediate technology into a worldwide movement starting in the mid-1960s.

Essentially, intermediate technology stands half-way between traditional and modern technology, obviously all of them relative notions. Schumacher used the term intermediate technology to refer to a technology that would provide jobs ("workplaces") for an amount of capital somewhere between the very large investment required for the advanced technology of the industrialized world and the very small investment used in traditional societies. Schumacher called these £1,000-per-workplace technologies in the advanced countries compared to £1-technologies in traditional developing societies. And what has been happening is that the £1,000-technologies have been eliminating traditional village workplaces at an alarming rate without providing any alternative employment for the large numbers who thus lose their livelihood. While the traditional technologies are grossly inefficient and wasteful of skill and resources, the modern £1,000-technologies are often too complex and always too expensive for these societies. Such technologies are meant for the rich and the powerful.

Thus, to Schumacher an intermediate level of investment in the £100-technologies would achieve a number of desirable objectives in these traditional societies. To wit, these intermediate technologies would be more productive than the indigenous technologies of the £1 variety but also cheaper and more manageable than sophisticated large-scale modern technologies. They would create more jobs and generate economic activity in short order that was not only suitable to local conditions but also fitted the existing levels of education, skills, know-how, and business experience. In short, Schumacher saw intermediate technology as having special application in the development process to meet the needs of the poor. Schumacher's "heir" in the intermediate technology movement gives the following example of IT:

> A small-scale brickwork producing 10,000 bricks a week instead of 1 million with advanced equipment but where capital costs are £400 per workplace instead of £40,000 in a large, modern brickworks. The smaller plant also saves fuel costs by air-drying bricks before firing, and local production virtually eliminates transportation costs. The final product of the smaller plant costs about half that of the capital-intensive, mechanized plant.[2]

According to Schumacher, intermediate technology

is characterized by the following criteria:

1. Job-generating capital investment must be created without requiring unattainable levels of capital formation or imports. Hence, intermediate technology is sometimes termed <u>light-capital technology</u> (LCT) characterized primarily by its low capital requirements and the small size of the investment needed to create a job. The term <u>capital-saving technology</u> (CST) is also used.

2. Workplaces must be created in the countryside where the people now reside, i.e., must be decentralized, rather than concentrated in metropolitan areas. That is, workplaces must be established where there is major unemployment so that jobs may be available to those who need them most and villagers are not forced into urban migration, swelling the masses of unemployed in the cities.

3. Production methods employed must be relatively simple and small-scale so that the need for advanced management and/or production skills may be minimized and fall within the capacity of users, so that work may give people self-reliance and self-esteem, and so that technology may serve people and not vice versa.

4. Production must originate mainly out of local resources and primarily for local use and benefit.[3]

Because for Schumacher intermediate technology set out to solve the problem of rural unemployment and urban drift and was thus defined in a particular way, the recipe that it offered enshrined certain value judgments (e.g., that any work is better than no work because the human need to be usefully employed is paramount or that technology should be non-violent to society or nature and nondisruptive of traditional lifestyles) which would have been different if the problem had been defined in terms of different values--for instance, making the society's aggregate production or modernization the top priority. In the latter case, the solution would have called for major imports of advanced Western technology

43

regardless of human, societal, or environmental disruptions.

The point is that a technical solution to an essentially social problem tends to build into itself the values which are held to be important by the way the problem is defined. If the problem is posed differently, other values are held to be important and a different solution enbodying those values will follow.

Now, for intermediate (or any other) technology to be successful, society has to be organized in a certain way. In the case of intermediate technology a decentralized, regional organization is a prerequisite with corresponding economic, social, and political implications. But the choice of technology is at least ultimately if not proximately a political decision determined by those in power, not an objective "neutral" or "scientific" one to determine "who gets what, when, and how," which is the essence of politics.

Albeit, no matter in what sense it is used, one must recognize that intermediate technology is a relative term which varies in time and space. For the same technology which may rate as intermediate in one society (say, an ox-drawn plow in a society where hand-operated hoes are more common) may be viewed as traditional elsewhere (because here it is the more sophisticated tractors that are considered intermediate, while agricultural combines rate as advanced technology).

Appropriate Technology

The terms intermediate technology and appropriate technology (or AT), also a generic term to describe a wide range of technologies, are often used interchangeably. Yet, the two are not the same. First, historically, the concept of intermediateness preceded that of appropriateness and took shape before the disenchantment felt around the 1970s in the Third World with Western technology transfer that reflected the factor intensities, values, and needs of modern societies rather than those of traditional societies.[4] Second, while both these types of alternative technologies emphasize the use of labor-intensive and capital-saving means, appropriate technology is more forthright in adding to this essentially quantitative base stressing techniques a much broader social and cultural dimension having to do with what kind of life

or society people want and how they go about organizing themselves to acquire it.

Accordingly, appropriate technology is characterized by at least one but usually several of the following features: Low investment cost per workplace,
low capital investment per unit of output, low cost of
final product, high potential for employment, but also
sparing use of natural resources, organizatonal simplicity, and high adaptability to a particular social or
cultural environment.[5] That is why Schumacher often
called it "appropriate technology from a human point
of view."

Appropriate technology relates to all aspects of
community needs and is a total approach to development. It is not only concerned with how many things
or jobs are created but also with the preservation of
lifestyles meaningful to the technology-receiving society and in keeping with such values as self-sufficiency, self-reliance, harmonious cooperation with
others, integration with the surroundings, decentralization, accountability, and participation. In short,
appropriate technology is an approach rather than a
specific level or package of techniques or tools. It
has much more of a human face than intermediate technology since, being more qualitative, appropriate
technology is more intimately tied to questions involving the goals and ideals of people.

It is then readily apparent why the term appropriate technology does not have exact parameters, has
no distinctive semantic content of its own, but rather
that it is situational, relational, and dynamic. It
is always necessary to specify appropriate to whom and
to what--to which country or region, to what kind of
development, for which human or environmental needs--
in short, appropriate to what value system.

For appropriateness is little more than a useful
tautology which sensitizes us to the existence of a
wide array of technologies among which the one best
suited to particular circumstances can, at least potentially, be identified or devised.[6]

Except in free market economies where consumer
demand may set limits to the use of appropriate technology, these circumstances often leave open the question of who should determine the criteria of appropriateness (the recipient firm, the supplier firm,
village, host government, or international

45

organization). This ambivalence explains in part why different organizations have their own working definition of appropriate technology.

For instance, the World Bank distinguishes among four dimensions of appropriateness: Appropriateness of goal (does the technology support the objectives or ideals of change?); appropriateness of product or service (is the final product or service useful, acceptable, and affordable?); appropriateness of process (is the use of inputs, degree of local participation, extent of local decision-making suitable?); and environmental and cultural appropriateness (does the transferred good or service blend with the physical or human-values environment?). This will suggest that appropriateness is not only a matter of degree but that often the same technology may have both appropriate and inappropriate features with reference to a given situation and that any choice involves a trade-off between the desirable and undesirable aspects of that technology depending on the value criteria applied and local constraints and opportunities.

TRANET (Transnational Network for Appropriate/ Alternative Technologies) defines appropriate technology as having the following characteristics: It is low-cost, easily maintained, uses local materials, protects the environment, is resource-conserving, fits into established cultural patterns, and increases the well-being and the dignity of people.

VITA (Volunteers in International Technical Assistance) considers appropriate technology as filling a real need expressed directly by the recipient and beneficiary of the technology, increasing production, labor- rather than capital-inventive, making maximum use of local available materials, resources, and skills, and being compatible with local traditions and customs.[7]

Jequier, who noted that appropriate technology represents "the social and cultural dimension of innovation,"[8] has identified some of the less widely used criteria of appropriate technology. These are systems-independence (the extent to which a given technology can do without supporting services), its image of modernity (the user's perception of having the latest and the "best"), individual versus collective technology, cost (economic, ideological, moral, social), risk (internal risk relating to inherent design characteristics, but also external risk relating to the ways in which a technology fits into the local production system, available supporting services, and local culture, with the two types of risk tending to operate in a synergistic way), evolutionary (dynamic) capacity for change, and single-purpose versus multipurpose technology.[9]

Schumacher, the founder of the intermediate-appropriate technology movement, eventually wished every transfer of technology to be subjected to certain searching questions such as: Does the contemplated technology transfer conserve energy? Does it enhance environmental quality? Does it humanize the work environment? Does it help the poor? From this perspective, to apply merely an economic gauge to the outcome of technology transfer (even though a dollar criterion also implies some value-related framework) can be misleading because growth in national income or per capita income can be accompanied by rising unemployment, social tensions, misery, and frustration.[10]

Trade-Offs

Amulya K. N. Reddy of the Indian Institute of Science in Bangalore developed the concept of environmentally sound and appropriate technology (ESAT), a form of appropriate technology particularly well adapted to the local economic and sociocultural environment in terms of:

1. Satisfaction of basic human needs;

2. Resource (including human resource) development;

3. Social development;

4. Cultural development;

5. Environmental development.

He divided each of these into further subcriteria. But he concluded that the best that can be done is to weigh the criteria and settle for trade-offs among them, hopefully to develop new techniques that allow as many criteria as possible to be satisfied simultaneously. Reddy wrote:

It is almost certain that, from the standpoint of the list of criteria proposed..., few current technologies are perfectly

47

environmentally sound and appropriate.
It is only a matter of some technologies
being more environmentally sound and appropriate than others. But, the revela-
tion of the gap between the ideal and the
actual provides the motivation for attempting to narrow the gap, i.e., for increasing
the environmental soundness and appropriateness of technologies....[11]

Other variations of appropriate technology include community technology (CT), a term used by the
American counterculture to refer to small-scale,
simple infrastructure specifically tailored to the
needs and capabilities of small urban or rural communities.[12] Community technology seeks to foster community participation in decision-making regarding such
projects as small-scale cooperative industrial activities or decentralized water supply and waste disposal
systems.

On its part, socially appropriate technology (SAT)
stresses technology with beneficial effects on employment, income distribution, work satisfaction, health,
and social relations. There are several other types
of appropriate technology, emphasizing in turn the
soft, nonviolent, adaptive, and progressive aspects of
technology.[13]

Appropriate Technology Movement

The major contribution of the appropriate technology movement, which relates to these various alternative technologies, has been to show that all technical
choices are value-laden, whether the decision-maker
realizes it or not. And no longer do users routinely
opt for the most advanced technology available without
thought to the myriad consequences of such transfers.
Albeit, the movement also has its critics.

Some have seen it as a recipe for continuing
underdevelopment in the LDCs.[14] Others say that the
movement perpetuates gender-related (to feminists,
male-dominated) traditional values so that even appropriate technology relegates women to such "soft" work
as agriculture, secretarial positions, and household
chores while the major technical projects and decision-making are still reserved to men.[15] Still others
refute the alleged benefits of small being always
beautiful or possible; of rural self-sufficiency being
always idyllic; of decentralization being always

48

practicable or desirable; and on other grounds.[16]

Inappropriate Technology

Both those who choose technology, whether for production or for social development, and those who consume it may be incompetent to make technical choices or may be uninformed about alternative choices. Specifically in the case of a technology supplier, for a number of reasons (multinational corporations are often ascribed self-serving ends) it may wish to impose its dominant technology on its weaker trading partner. In the case of a technology recipient, it may lack the means to carry out its own research and development, innovate, choose the most appropriate technology, adapt imported technology to its own requirements, or may also be swayed by a self-serving domestic elite which often monopolizes resources and political power.

As a result of these and other reasons,[17] a technology is often adopted as is, rather than adapted to the very different economic, social, cultural, psychological, or other needs of the recipient society or at times simply different environmental or climatic conditions. Such inappropriate technology, which has less than an optimal socioeconomic cost-benefit ratio (again, a relative concept involving definitional problems), has been compared to a lock and key analogy as follows:

A technology that is energy-intensive, capital-intensive, labor-extensive, research-intensive and organization-intensive transferred to a place (a district in the same country or another country) which is short of energy supply, has little capital, abundant labor, very little research capacity and is low on managerial skills, is like a key put into a lock, but with the opposite profile of the one needed to open it. Of the two possibilities, changing the key or changing the lock, it looks as if the development ideologies of the last decades have allowed the holders of the keys to oblige the others to change the locks rather than vice versa. In other words, a new "lock" has been created for the key by concentrating in some place

49

sufficient nature (particularly energy),
capital, research capacity and organization capacity--often by importing all
four on a grant or loan basis--to provide
a factor-environment that can meet the demands of the technology. In this manner
an enclave is born, adapted to the imported technology, with a concentration of
natural resources, capital, laboratories,
researchers, bureaucrats and capitalists,
not drawing on the workers of the land,
who are usually peasants. In short, the
technology reproduces itself by forcing
the economic environment to adapt. This
has well-known social consequences: The
concentration of wealth in the capitals
and other "growth-poles," surrounded by a
vast periphery totally unable to sustain
a modern technology because of the factor-
concentration in the growth-poles.[18]

The way the "lock" has been changed to fit the
"key" is evident from a case study involving the hotel
industry in Jamaica.

The author describes how the developed countries had transferred to this developing
Caribbean island an advanced Western technology intended to reproduce a standard of
living congenial to the relatively well-to-
do Americans, Canadians, and Britons who
represent most of the visiting tourists.
Thus, the island is known for its jet travel
and luxury hotel accommodations and spends
much of its foreign currency importing related technology and consumption goods. Yet,
what this Third World society needs is not
French-style bidets for its fancy hotels but
rather simple community-like housing and
local transportation modes as well as agricultural improvements.[19]

The inappropriateness of tourist transfer has been
stigmatized further for its clash with basic cultural
and other values of those developing countries that
emphasize it. To illustrate this point, Goulet quotes
the former prime minister of St. Vincent, another
Caribbean island, who wrote some years back: "As Pre-
mier of my state, you will pardon me, I hope, if I
appear not too anxious to grab the easiest dollar.
The tourist dollar alone, unrestricted, is not worth

the devastation of my people. A country where the people have lost their soul is no longer a country--and not worth visiting!"[20]

Admittedly, the earlier discussion of appropriate technology will have suggested that it is not always easy to draw the line between appropriate versus inappropriate technology. Consider the following example:

> An industrialist wished to introduce breweries into Botswana, southern Africa. Lager beer was already being imported from neighboring South Africa, and there was a counterproposal to build a large modern brewery in the capital of Botswana. The industrialist advocated the building of several small-scale, village-based breweries to meet two important objectives of intermediate technology mentioned earlier: (1) To create job opportunities at the village level; and (2) to reduce urban drift. The effects of this alternative rather than building one large modern brewery in the capital was that distribution and supply of beer throughout the country could be made easier and hence beer would be more readily accessible.[21]

The question can then be raised whether, even though the village-based breweries involved intermediate technology, they were in fact appropriate. The answer is not self-evident because the easier accessibility of beer at the village level had considerable social implications--decreasing standards of nutrition because money was now being spent on beer rather than on food, increasing crime rate because of the effect of alcohol, and the consequent neglect of traditional work. On the face of it, then, and in the last analysis, intermediate technology in this case was not appropriate.

Or think about this second example:

> About 50 years ago white men first penetrated into the valleys of the interior of what is today Papua New Guinea, islands north of Queensland, Australia. In order to create goodwill among the natives, the whites gave them metal axes as presents. This inadvertent introduction of intermediate technology--metal axes to replace the

51

existing stone axes used in cutting trees,
working gardens, etc.--while not changing
the organization or control of village life,
led to an unexpected result whose outcome
the author of this case study summarizes as
follows:

"In the first place the introduction
of the new steel technology raised the potential supply of goods of all kinds, since
it set free time that could have been used
to make any kind of good. No more subsistence goods were produced, since demand for
them was stable given the existing role
structure of society. Time was spent in
efforts to increase the power of each individual and the group. Some of these efforts
took the form of fighting to obtain power;
some were efforts to obtain power through
the increased use of valuables."[22]

Again, the transfer of intermediate technology in this
case was not necessarily appropriate any more than it
was in the case of Cretan-type windmills which were
introduced into northern Ethiopia. For the canvas
used in their sails was removed by village girls as
skirt material until replaced by another less suitable
for wear.[23]

Studies have indicated that in developing countries a major failure in the planning process occurs
if alternative investment strategies are not maintained as viable options during the entire evaluation
stage, and failure is inevitable if alternative strategies are never entertained in the first place.[24]
This is a pronounced experience and may occur because

(1) It is easier from a bureaucrat's viewpoint to design, construct, administer,
regulate, or control a few large capital-
intensive projects than many small labor-
intensive ones.

(2) Western-trained engineers and economists
have a natural bias toward "efficiency," often regardless of social or other environmental considerations. That is, where engineering and/or economic values are unques-
tioned, efficient solutions are defined
without reference to social and cultural
externalities which are thus excluded from

52

cost-benefit calculations.

Consider the case of a North African country at
one time producing leather sandals for local use.

Policy-makers decided to lower production
costs by importing two plastic molding injection machines to turn out plastic sandals. The installation of the highly
automated plant led to the unemployment of
5,000 leather shoemakers and of others in
ancillary industries (such as glue-making)
since only 40 injection mold operators were
now needed. There was also a decline in
domestic income, in foreign currency reserves because of the need to import poly-
vinyl chloride to produce the plastic, and
other consequences.[25]

In another case study of footwear-
making technology transfer to Ghana and
Ethiopia and of sugar-production technology
to Ghana, the authors concluded that "the
explanation of imprudent decisions lies in
the malign influence of the engineer and
in the conceit of the economist. Less
provocatively put, the investment decision,
as it is widely made, is an engineering
one subject to a broad economic constraint.
A decision is taken, for example, to establish a plant of some given productive capacity in a developing country. Engineers
trained according to developed country curricula are asked to design the plant. They
produce blueprints for a limited number of
alternatives, each of which is a variant of
current 'best practice' technologies. The
alternatives are submitted to economic...
scrutiny, the most attractive chosen, and
another capital-intensive, technologically
inappropriate plant is established."[26]

This is what has prompted Goulet to observe that "what
is needed is a new breed of technicians and engineers
who, if they are not themselves philosophers, are
willing to trust the philosophical judgment of common
citizens in the political arena."[27]

It does not seem possible to generalize about the
types of decision-maker who tend more than others to
make imprudent decisions. For example:

53

A multinational corporation using engineers from a developed country was reported to have been responsible on two occasions for its inappropriate choice of the most capital-intensive type of can-making technique in Thailand.[28] Here, the conclusion was that since no particular categories of decision-makers could be shown to be more imprudent than others, the origins of imprudent decisions seemingly had to be found among influences that cushion the impact of prevailing prices and costs (e.g., of labor, capital, raw materials). Limited knowledge of can-making in Thailand, risk avoidance, and the desire for system compatibility among multiplant firms (the multinational corporation's plants at home and in Thailand) were determinant in the technology selection process.

In the case of a U.S. Agency for International Development-financed Djakarta-Bogor highway in Indonesia, the choice by U.S. AID contract engineers of an American-style, limited access, four-lane highway to be built with modern engineering equipment and maintained in the same fashion ignored more labor-intensive alternatives less familiar to the American designing engineers.[29]

Why should the local government authorities have gone along with a technology which local civil servants may, on their part, have considered as inappropriate? If foreign aid is available only for that choice for whatever parochial, even self-serving, economic or political reasons, the recipient government knows that this may be the only way to receive the desired foreign assistance. Indeed, officials may tailor their own program to the technology they consider most likely to attract foreign grants or loans or other transfers.

Or, for a number of reasons (corruption, hyperinflation, etc.) civil servants holding key positions in developing countries have often found it more remunerative to work with large, capital-intensive projects, especially those with a substantial import component, than the smaller, more indigenous labor-intensive ones since the capacity to generate side payments or other forms of perquisites is greater in the former case. Hence, these officials are

disinclined to search out and consider viable labor-
intensive projects and may endorse inappropriate ones,
especially as many Third World engineers and policy-
makers are trained in the West where appropriate technology is not generally studied.

Consider the case of the transfer of tractors to
Pakistan.

In the late 1960s that country obtained a
World Bank loan to purchase 18,000 tractors,
made available to large landholders on gen-
erous credit terms. The result was that the
recipients increased their output and income
substantially following the introduction of
the tractors. But by the early 1970s, while
the tractorized farms more than doubled in
size, many smallholders and tenant farmers
were forced off their plots and the use of
hired labor dropped by some 40 percent as
each tractor led to the net loss of about
five jobs. Too, the tractors were found to
have no effect on crop yields or the number
of crops grown each year. In short, as with
the green revolution, the transfer of tractors to Pakistan at that time was discovered
to have socially regressive results.[30]

Ponder, too, food technology transfers with reference to appropriateness or, more likely, otherwise.

In one instance, an effort was made to introduce a hydraulic palm oil press in Africa. The change agents communicated the
information to males in the community, ig-
noring traditional sex roles. The women
who customarily pressed the seeds rejected
the press because it destroyed the byproducts, very important to them. In another
instance in Ghana, a new type of oven was
introduced for smoking fish which required
putting rods through their eyes. Since eye-
less fish were unacceptable to the local
population the new technology was inappropriate. Or, in Asia a raised cooking stove
was introduced to protect the food from dirt
and animals. The stove was rejected because it mandated cooking in a standing
position, found uncomfortable compared to
the traditional stooping position.[31]

Other examples of the transfer of inappropriate technology in food processing include lager beer brewing techniques from the West to produce beer with excessive standards of clarity but at the cost of large investments in equipment and raw materials and the generation of little employment. Or, the making of bread with sifted instead of wholewheat maize even though the latter technique is small scale, more capital-saving and labor-intensive, and the product more nutritious.[32]

On a larger scale and for different reasons, the Aswan High Dam in Egypt, completed with Soviet technical and financial assistance in the late 1950s and 1960s, was recently cited as another example of inappropriate technology. The dam's gigantic size is matched by the monumental problem that it was designed to solve--to help feed a population of well over 40 million on a cultivated area of between 3 and 4 percent of the country's territory because of prevailing desert conditions and which is yearly being reduced under the pressure of urbanization, industrialization, and road-building. The following are excerpts from comments made by Edgard Pisani, former French minister and now a member of the Commission of the European Communities in charge of development:

> One will thus understand why the Aswan High Dam should have appeared and may still appear as an absolute necessity. Accordingly, this eighth masterpiece in the history of humanity, this eighth wonder of the world was built and the greatest miracles were expected from it. The Aswan High Dam creates the possibility of irrigating... 8,000,000 acres. At present the rate at which these irrigable areas are being effectively put under irrigation is in the order of 20,000 hectares [about 40,000 acres] a year. At this rate, eventually 200 years would be necessary to fully utilize the hydraulic and irrigation capacity of the Aswan High Dam; that is, by the time the last acre is irrigated, the dam will no longer be in existence. Thus, the first problem is that an extraordinary tool was placed into the hands of a country...but the country's social structures, its political structures, its administrative structures had not been readied to make possible the use of this extraordinary potential. An

instrument was made available to this country which it was not immediately able to use; in other words, preference was given to the tool instead of to the hand that was to employ the tool. The result is that the rate of extending irrigation has been low.

But there is worse! This project has indeed completely disrupted the natural balances and I would like...to note that the silt, which the Bible represents as one of the major elements in Egypt's prosperity, is now trapped by the dam and cannot move downstream. In other words, whereas formerly the water was filled with mineral or organic matter, the water which now flows downstream is clear but is hard water which erodes the banks and bridge supports; but especially it is poor water which cannot enrich the fields around it, for it does not carry any organic element, any mineral. Egypt is thus obliged to purchase fertilizers.

Furthermore, it so happens that the High Dam was built at a time when electric pumps were being generalized on the market. Thus, instead of the water flowing naturally and allowed to flood the land and drain off at a natural rate [under the basic irrigation system], it now remains at a fairly low level and must be pumped. Since it is no longer necessary to exert physical effort to pump water [manually] and now that it is sufficient to hook up the [electric] pump, there is a tendency to overirrigate...while it is necessary to add fertilizers. Since there is no physical limit to pumping, too much water is used and the water gradually seeps into the ground and reaches the layer of salty water ...which, through capillary action, causes the salinity to rise to the surface. Overall, the Aswan High Dam is negative.[33]

Albeit, some people emphasize the positive benefits of this macro project, pointing out that the pertinent question is not whether the High Dam should have been built but rather what measures should be taken to sustain it over the long term.[34]

It should be stressed that inappropriate just like appropriate technology is also situational, relational, and dynamic and that a given technology transfer may be appropriate at one point but inappropriate at another, following a change of circumstances. Consider this case of domestic technology transfer suggesting economic reasons for inappropriateness as a function of time:

In late 1981, the McLouth Steel Corporation was facing bankruptcy because it had acquired some of the best technology in the business! Indeed, the Detroit-based steelmaker had been one of the first to install the innovative basic oxygen furnace in 1954. By the early 1970s, it was the first to process all of its steel through the efficient continuous casting method.

However, McLouth depended for 75 percent of its orders on General Motors. When the latter cut back on its purchases because of the slump in the automobile industry in the early 1980s, McLouth laid off half its personnel and ran its costly equipment at much less than capacity. Still, as a former technical director noted: "That kind of equipment does not take kindly to running at half speed. You just can't scale down that easily." In other words, the "best" equipment had made McLouth less flexible and its operating costs spiraled as demand for its flat rolled steel ebbed.[35]

The following mini example illustrates how the ignorance of local mores and customs nearly resulted in the transfer of inappropriate communications technology:

In order to depict a woman who has forgotten to take a contraceptive pill and slept through the night, an initial illustration prepared by the United States Program for the Introduction and Adaptation of Contraceptive Technology in Health (PIACT) represented the subject with her hair loose calmly sleeping next to a glass of water and the forgotten pill. After consulting with Bangladeshi women from the intended target audience, it was learned that to them this illustration conveyed the idea

58

that taking the pill had killed the subject, as Bengali women normally sleep with
their hair tied back. The composition was
then changed and a potentially serious mis-
understanding averted.[36]

Since it is PIACT's goal to develop culturally appropriate print materials for non-readers and since it
goes about doing this by involving and interacting
with representatives of the target audience, it would
be interesting to speculate how the inappropriate
graphics were prepared in the first place. Perhaps
the artist preparing the material was unsuspecting of
the symbolic significance of hair-dos in an exotic
culture or did not feel that it was relevant to the
specific message involved. One should also be mindful
of the difficulty in selecting the right picture for
non-readers whose perceptions are largely unknown.

At times, other practical reasons may preclude
the use of appropriate technology. The necessary
machinery, equipment, or spare parts may no longer be
produced. Or, as was indicated earlier, the techno-
logy-importing countries may not be interested in what
they consider to be second-rate systems. Or, the suppliers of technology may enjoy monopoly market power
and decide unilaterally which form is most "appropriate" even though this may not be so. Or, market
demand may prompt the transfer of inappropriate products (for instance, for imported cola drinks in
countries with very low per capita incomes and poor
diets).

For these and other reasons, there are those who
view all technology transfer between societies of different levels to be inappropriate. A scientist expressed this kind of advocacy as follows:

Transferring technology from industrial to
developing countries is a form of neo-
colonialism we are trying to overcome.
People at the local level should participate in the choice, design, construction,
ownership and control of technologies meant
to help them. Technologies designed for
them from the outside even more than technologies transferred to them which they do
not understand are destructive of their
culture, economics and ecologies, and their
dignity. What we are concerned with

transferring is the spirit of technological innovation.[37]

Summation

This chapter broadly dichotomizes as between appropriate and inappropriate technology in order to emphasize that only a technology that is compatible with the recipient's physical, legal, economic, but especially sociocultural and psychological environment is likely to be suitable. In practice, however, the terms appropriate and inappropriate are situational, relational, and dynamic making the dichotomy even harder to establish. Additionally, various kinds of trade-offs, or sacrifices, are often involved since it is not possible to maximize all aspects of a transfer simultaneously. Such ambivalence makes the dichotomy even harder to establish.

In the meantime, the widespread transfer of advanced, modern technology, the obverse rather than the reverse of alternative technologies such as appropriate technology, continues. This is because of such factors as prestige, the lack of evaluational skills especially by technology recipients, vested interests and other self-serving reasons on both sides of the transfer process, as well as the fairly prevalent belief that appropriate technology is second best whereas the goals of technology transfer such as development call for state-of-the-art technology.

FOOTNOTES

1. E. F. Schumacher, Small Is Beautiful: Economics As If People Mattered (New York: Harper and Row, 1973).

2. George McRobie, Small Is Possible (New York: Harper and Row, 1981), p. 42.

3. Schumacher, Small Is Beautiful, p. 146.

4. As mentioned earlier, this does not mean that traditional societies--both the elites and the masses --have been averse to adopting Western technology and with it Western lifestyles when these are completely out of kilter with what their culture is about.

5. Nicolas Jequier and Gerard Blanc, The World of Appropriate Technology: A Quantitative Analysis

(Paris: Organization for Economic Cooperation and Development, 1983), p. 10.

6. Gustav Ranis, "Appropriate Technology: Obstacles and Opportunities," in Samuel M. Rosenblatt, ed., Technology and Economic Development: A Realistic Perspective (Boulder, Colo.: Westview Press, 1979), p. 25.

7. Malcolm C. Bourne, "What Is Appropriate/Intermediate Food Technology?" Food Technology (Chicago), vol. 32, no. 4 (April 1978), pp. 77-80.

8. Jequier, Technology Transfer and Appropriate Technology, p. 19. See also United Nations Industrial Development Organization, Conceptual and Policy Framework for Appropriate Industrial Technology (Monographs on Appropriate Industrial Technology No. 1) (Vienna: UNIDO, 1979), p. 3.

9. Nicolas Jequier, "Appropriate Technology: Some Criteria," in A. S. Bhalla, ed., Towards Global Action for Appropriate Technology (Oxford: Pergamon Press, 1979), pp. 1-22.

10. George McRobie, quoting a speech by E. F. Schumacher delivered in Caux, Switzerland, on September 3, 1977, on the eve of Schumacher's death--McRobie, Small Is Possible, pp. 1, 27. The use of the binomial "intermediate/appropriate" does not contradict the earlier distinction between the two concepts because, over time, they have tended to converge as Schumacher's parameters themselves will suggest. Schumacher's Intermediate Technology Development Group in London eventually changed the name of its newsletter from IT Bulletin to Appropriate Technology Journal. One author has solved this definitional ambivalence by referring to both as "modest technologies"-- Roberto Vacca, Modest Technologies for a Complicated World (Oxford: Pergamon Press, 1980).

11. A. K. N. Reddy, Technology, Development and the Environment: A Reappraisal (Nairobi, Kenya: United Nations Environmental Programme, 1979), pp. 28, 34.

12. Karl Hess, Community Technology (New York: Harper and Row, 1979).

13. Yvonne L. Hunter, "Introduction to E. F. Schumacher," in Dorf and Hunter, eds., Appropriate Visions, pp. 65-67.

14. Arghiris Emmanuel, Appropriate or Underdeveloped Technology? (translated from the French by Timothy E. A. Benjamin) (New York: Wiley, 1982).

15. Judy Smith, "Women and Appropriate Technology: A Feminist Approach," in Jan Zimmerman, ed., The Technological Woman: Interfacing with Tomorrow (New York: Praeger Publishers, 1983), pp. 66-67.

16. Malcolm Hollick, "The Appropriate Technology Movement and Its Literature: A Retrospective," Technology in Society, vol. 4, no. 3 (1982), especially pp. 213 and 228.

17. Discussed extensively in Frances Stewart, Technology and Underdevelopment, 2d ed. (London: Macmillan, 1978), ch. 3 and 4. This British economist argues that appropriate technology strategy for development happens to coincide with conventional economic rationality.

18. Johan Galtung, Development, Environment, and Technology: Toward a Technology for Self-Reliance (document TD/B/C.6/23/Rev. 1) (Geneva: United Nations Conference on Trade and Development, 1979), p. 19.

19. John F. E. Ohiorhenuan, Social, Environmental, and Economic Aspects of Technology Transfer in Jamaican Tourism (UNEP/UNCTAD TD/B.C6/49) (Geneva: United Nations Conference on Trade and Development, November 28, 1979).

20. Remarks of James F. Mitchell, cited in Jerry Kirshenbaum, "To Hell with Paradise," Sports Illustrated, May 21, 1973, p. 50. A subsequent rebuttal of this position was voiced by a Greek-born professor of sociology at the University of Paris who opined that "cultural authenticity is also the tourist picturesqueness of underdevelopment....Humanity is neither a zoo nor a museum of the anthropologically exotic." Emmanuel, Appropriate or Underdeveloped Technolgoy?, p. 106.

21. See P. van Rensburg, "Small and Simple Breweries for Rural Africa," The Brewer, March 1975, quoted in Lipscombe and Williams, Are Science and Technology Neutral?, p. 35.

22. R. F. Salisbury, From Stone to Steel: Economic Consequences of a Technological Change in New Guinea (Melbourne: Melbourne University Press, 1962), in

ibid., p. 36.

23. Peter D. Dunn, Appropriate Technology: Technology with a Human Face (New York: Schocken Books,
1978), pp. 8-9.

24. These comments are based on C. Peter Timmer, John
W. Thomas, Louis T. Wells, Jr., and David Morawetz,
eds., The Choice of Technology in Developing Countries:
Some Cautionary Tales, Harvard Studies in Internation-
al Affairs No. 12 (Cambridge: Harvard University
Press, 1975).

25. Dunn, Appropriate Technology, pp. 16-17.

26. James Pickett, D. J. C. Forsyth, and N. S. McBain,
"The Choice of Technology, Economic Efficiency, and
Employment in Developing Countries," World Development
(Oxford, England), vol. 2, no. 3 (March 1974), p. 51.

27. Goulet, The Uncertain Promise, p. 28.

28. C. Cooper, R. Kaplinsky, R. Bell, and W. Satyarak-
wit, "Choice of Techniques for Can-Making in Kenya,
Tanzania, and Thailand," in A. S. Bhalla, ed., Technology and Employment in Industry (Geneva: International Labor Office, 1975), p. 117.

29. Timmer, et al., pp. 12-13.

30. Colin Norman, The God That Limps: Science and
Technology in the Eighties (New York: Norton, 1981),
pp. 155-156.

31. Richard B. Pollnac, "Sociocultural Factors Influ-
encing Success of Intermediate Food Technology Programs," Food Technology, vol. 32, no. 4 (April 1978),
p. 90.

32. Christopher G. Baron, "Appropriate Technology,
Employment and Basic Needs in Arab Countries with
Special Reference to the Food Industries," in Antoine
B. Zahlan and Rosemarie Said Zahlan, eds., Technology
Transfer and Change in the Arab World (New York: Per-
gamon Press, 1978), pp. 387-388.

33. Edgard Pisani, "The Third World's Struggle Against
Hunger and Development," Ensemble (Lille, France),
vol. 40 (new series), no. 1 (March 1983), p. 8 (in
French). The social costs of high dams in general
were discussed in Philip Shabecoff, "Actual Price of

High Dams Also Includes Social Costs," <u>The New York Times</u>, July 10, 1983, p. E22.

34. Margaret R. Biswas and Asit K. Biswas, "Environment Implications of Development for the Third World," in Barbara A. Lucas and Stephen Freedman, eds., <u>Technology Choice and Change in Developing Countries: Internal and External Constraints</u> (Dublin, Ireland: Tycooly International, 1983), p. 125.

35. Lydia Chavez, "Pitfalls of New Technology," <u>The New York Times</u>, December 17, 1981, pp. D1 and D6.

36. Joan Haffey and Margot Zimmerman, <u>Overseas Development</u> (Washington, D.C.), October 1983, p. 14.

37. Personal communication from Dr. William N. Ellis of TRANET (Transnational Network for Appropriate/Alternative Technologies) to the author, January 11, 1982. This view epitomizes that of a slew of "radical" technologists such as Godfrey Boyle, Norman Colin, Robert J. Congdon, Ken Darrow, Peter Harper, Ivan D. Illich, Jack and Nancy Todd, and others. Some of their representative works are included in the Bibliography.

CHAPTER 4

PROBLEMS IN TECHNOLOGY TRANSFER

The effort to pinpoint a technology for a particular need involves, first, identifying the possible technological options and then, from among the possible, choosing the one that is most appropriate. This means that technology selection must take into account local technical, economic, labor, social, cultural, and psychological conditions in order to make the transfer congenial to the recipient's universe and thereby enhance the chances of ultimate absorption of that technology into the new setting. For while opportunities and constraints define the possible choices, values relating to the physical and human environments determine the "best" possible option.

Selecting and Assessing Technology

Selecting Technology

As a broad guideline and even though this may sound tautological, the technology selection process should identify one that produces the desired results in terms of its being useful and hence valuable in a given setting. And indeed, the earlier discussion of human values and culture (Chapter 2) suggested that the degree of resistance to change accompanying technology transfer depends on its compatibility with existing values. Yet, since more than one value is usually involved and these values often conflict, it was noted, selecting appropriate technology means in effect establishing an order of priorities among different and often incompatible values.

For instance, should environmental conservation be sacrificed on the altar of economic growth with the selection of highly polluting but industrially effective equipment? Should individual freedom or more equal income distribution through decentralization but less efficient production be traded off against a higher income achieved through centralization and more efficient methods? Should economic freedom such as labor union activity be countenanced even at the price of higher unemployment or less automation? Hard choices are often involved so that the selection of a given technology from among alternatives or between indigenous and transferred technology is problematic.

For every decision involves many factors relatively few of which are technical and relatively many of which are value-related. And the "right" answer in selecting technology--if it exists at all--will vary from case to case and from place to place and even from time to time in the same place (recall the example of the McLouth Steel Corporation in Chapter 3).

Here are some typical examples of the kinds of choices that have to be made in the technology selection process:

1. The use of prefabricated parts or bricklaying for building dwellings;

2. The building of a river embankment or a dam for a hydroelectric plant;

3. The production of steel in a traditional blast furnace or by direct reduction in a fluidized bed;

4. The use of natural or enriched uranium in a nuclear plant;

5. Formal education or on-the-job training of technicians;

5. The promotion of rural health through the use of advanced methods or the support of traditional medicine in the countryside.

Among cases that may lead to technological innovation, indigenously or through transfer, are:

1. The processing of indigenous raw materials whose characteristics make existing technologies unsuitable--e.g., the recovery of metals from native ores; the production of preserves, cosmetics, and chemical or pharmaceutical products from indigenous plants and animals.

2. The need to produce a given commodity on a smaller scale than existing technologies allow, but at costs that are comparable--for instance, 3 tons of nylon thread a month whereas normal plants of large firms produce 300 to 800 tons a month; or 50 radio sets a day instead of 250.

3. The solution of problems that are specific to a given country--for instance, an endemic sickness.

4. The avoidance of a dependent situation, that is, when only one technology is available and is under monopolistic control--e.g., the technology for producing nylon from vegetable oils was a response to the initial monopoly of the technology for producing nylon from petrochemicals.

One example of a problem requiring adaptation of the original technology to local conditions occurred when a polymer-spinning technology purchased in the United States encountered difficulties in operation because the lower boiling point of the solvent at the altitude of Mexico City (about 7,800 feet high) had not been taken into account.[1]

Or take the pulp and paper industries. These were based on European and American-Canadian technologies using fiber raw materials originating from conifers, in plentiful supply in these regions. Now, developing countries in warmer latitudes normally lack conifers but have other kinds of fiber raw materials--for instance, bamboo, straw, and bagasse (sugarcane residue) which can be used to produce pulp and paper through slightly different processes.[2] At times, however, the need may arise to develop a completely new technology. Again, this may involve choices.

Ergo, the most important prerequisite for making a suitable technological choice is knowledge of the available alternatives.[3]

Consider the following. Suppose there are three possible solutions to the problem of providing water to small farmers from aquifers near the surface:

1. Designing and testing a small hand- or pedal-powered pump to be used by one or two farmers and promoting a system to market and maintain such pumps;

2. Encouraging the installation of diesel-powered tubewells serving 50 or so farmers and insuring the equitable distribution of

67

the resulting water supply through coopera-
tives;

3. Encouraging entrepreneurs to hire out
truck-mounted pumping equipment by the hour
to individual farmers.[4]

Each of these approaches constitutes intermediate
technology. But the choice among them should depend
on careful overall assessment of local technoeconomic,
geographic, ecological, social, cultural, and psychological factors as well as the desired balance between
growth (or efficiency) and equity. Again, one of the
technologies is more likely to be suitable, to be
appropriate.

Sometimes, the experience of using a simple technology has an educational and stimulating effect on
wider community activities and is a catalyst for more
rapid social development. Unfortunately, however, developing countries often lack the market power and
skills to direct technology to their problems. The
selection of the right technology seeks to make a wide
range of basic improvements available to the maximum
number of people. For example:

The minimum medically acceptable standards
for water supply may be lower than most
specialists allow because one of the most
important benefits of an improved supply is
not its purity but the opportunity it offers
for better hygiene. This being so, the same
investment can be used to get a lower quality of water to more people.

Similarly, equity-oriented technology insuring
maximum social justice may also happen to be the most
appropriate with reference to other values. For example:

In the well-known case of the baby milk formula, the most basic technology--breast-
feeding--is not only the most equity-
oriented because, among other advantages,
even the poorest can afford it; but it is
also the soundest technology biologically
because of its immunizing and birth-spacing
characteristics.[5]

Or consider:

The solar pump developed by a French firm in cooperation with the University of Dakar, Senegal, uses a widely available source of energy--the sun--to provide villagers with a scarce but vitally important commodity, water. Although technically sophisticated, it blends rather well with the social environment of Third World countries like Mexico where it is widely used. The solar pump requires virtually no maintenance. It has a very long working life. It would thus appear to be a sound selection from a technical viewpoint and not violative of local sociocultural conditions. This would also be true of the solar coffee drier developed by Canada's Brace Research Institute for Colombia and solar distillers for the waterless villages of Haiti.[6]

The process of technological choice for an investment project and the engineering of the project require design engineers to interact with socioeconomic planners and evaluators and with entrepreneurial decision-makers. After selecting a technology, the designers provide the evaluators and decision-makers with estimates of the investment and operational costs involved. These estimates are used by the evaluators in a prefeasibility study in which they account for and (when necessary) forecast in socioeconomic terms the costs and benefits of the project. The results of the prefeasibility study may require that the designers make changes--minor or even functional--in the proposed technology.

When an overall technology has finally been agreed upon, the designers undertake its preliminary engineering. Costs are then refined. And chances are that a design produced under a carefully planned set of requirements, constraints, objectives, and opportunities and evaluated under socioeconomic criteria will be appropriate in the sense that it will accomplish the objectives of the project and be adjusted to its particular set of design requirements, constraints, and opportunities.[7] Figure 4.1 illustrates the process of technology selection and design. Among other things, references to ecological conditions and the recipient's psychological and social characteristics will be observed in the diagram.

This cost-benefit analysis suggests that some of the alternatives lying within the realm of feasibility

69

FIGURE 4.1

FIGURE 4.1

THE PROCESS OF TECHNOLOGY SELECTION AND DESIGN

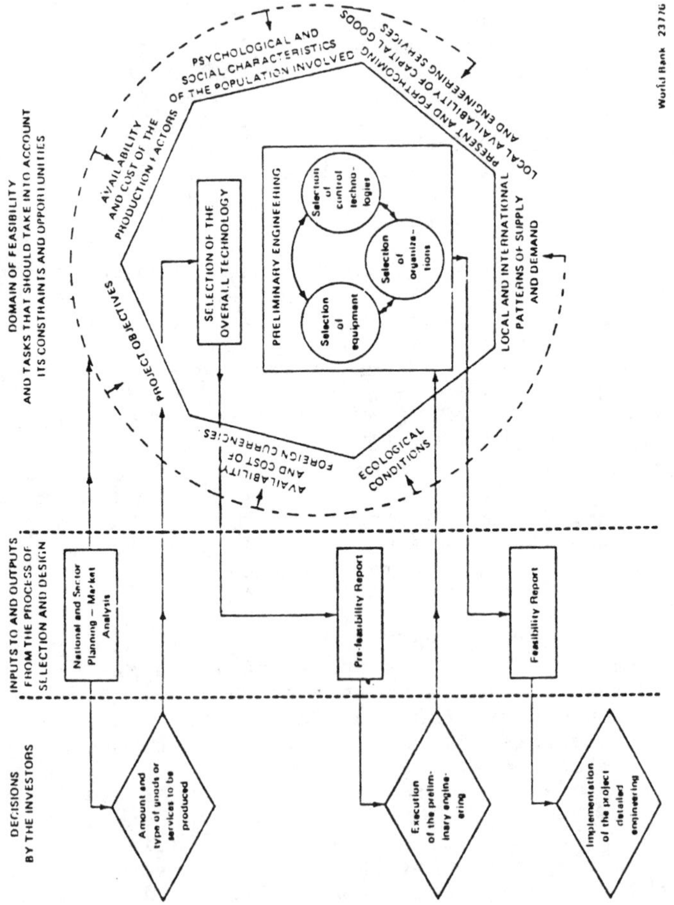

70

are rejected because they do not meet the minimum rates of return accepted by the designers or their organizations. Among the remaining options the designers make their final selection according to their set of values or that of the firm or organization to which they are accountable. For instance, some entrepreneurs may prefer to invest more capital but to deal with smaller numbers of workers to reduce personnel management problems if this is important to them. Other entrepreneurs may rather get the highest benefits but downplay the impact of a given process on the environment. Still others may select the alternative that would create more opportunities for the participation of workers and consumers in the design of the project and in the control of its accomplishments.

It should be noted that the value system of the designers and their organizations already played a role when the ranges for the different constraints and opportunities were established. These values are again factored in at the completion of the selection process when the minimum rates of return are established, shadow prices are calculated, and a technology is picked among the options that remain as economically viable after the cost-benefit analysis.

Assessing Technology

From the viewpoint of the supplier of technology, the transfer process is assessed in terms of whether it has met its objectives--to secure production factors at competitive prices; to keep or acquire a market; to gain access to regular sources of raw materials; to optimize the use of assets which may lack desirable alternative uses; to avoid ecological or labor constraints; or to maximize profits. Or else, the possessor of technology may withhold its transfer for any number of reasons discussed elsewhere.

In contrast, the host country or technology importing firm may assess the value of the incoming technology in terms of its penetration of local and/or international markets; its balance of payments or employment impact; the extent to which local capital or manpower participate in the transfer; the technical contributions made by the transferred technology to industrialization, development, or modernization; or the extent to which such technology may help solve any of the importer's other problems.[8]

But overall, given the difficulties of comparison

71

and of determining opportunity costs, differences in short- versus long-run effects, the cost of the assessment itself, and the fact that agencies often feel threatened by outside evaluations and are reluctant to engage in self-evaluations because of the potential of adverse criticism and the consequent drying up of funding, assessments are not always made or their results released.

In short, there is still a "need to develop low-cost, small-scale, highly participatory, internally-organized approaches" to evaluation.[9] In the meantime, then, the assessment of goals, especially relating to the transfer of appropriate technology which cannot be based on narrow, quantitative measurements, is often a matter of experience and personal hunches.

A number of examples by International Labor Office officials on major food processing activities make this evident.[10] Their 11 case studies involve rice milling, maize milling, bread baking, food grain storage, gari production (from cassava), coconut oil production, milk processing, sugarcane production, fish preservation, fruit and vegetable preservation, and beer brewing. These case studies indicate that technological alternatives, capital- and labor-intensive, small- and large-scale, are available to make products of varying qualities or refinement. While the studies confirm that technological choices exist in each process, generally applicable conclusions about the implications of the existence of choice do not emerge so easily.

Taking the case studies together, it cannot be claimed either that the most capital-intensive "modern" technology is necessarily the most prevalent or that the most labor-intensive "traditional" technology is always to be preferred. Although the editors are unsure about the feasibility of massive employment generation by the application of simpler but efficient processing technologies, they are clear that there is considerable scope for capital savings by using intermediate techniques rather than imported turnkey technology. Keddie and Cleghorn also note the potentially serious social problems which can arise in switching to the "optimal" technology. For instance, in rice milling, although the most labor-intensive technique, hand pounding, is not economical, the social consequence of promoting small-scale, mechanized techniques would be serious unemployment, especially in the Asian context.

Other noneconomic factors noted specifically in
connection with another case study on bread baking in
Kenya, such as cultural compatibility, environmental
effects, human enrichment, and the satisfaction of
basic needs, reinforce the notion that while cost
minimization may be a desirable objective, it is not
the only criterion of appropriateness in all circumstances. In this case:

> Native buildings housing bakeries (espec-
> ially in rural areas) have a lower imported
> building material content and require more
> labor-intensive construction techniques.
> Or, smaller bakeries reinforce the strong
> tradition of cooperative labor (in contrast
> to hierarchical relationships mandated by
> task specialization in large bakeries).[11]

Highlighting the difficulty of choosing which is
the appropriate technology in this instance, Kaplinsky
writes:

> If there were no trade-offs, it would be
> easy to decide which category of bakery is
> more socially appropriate. But there are
> certain trade-offs: Small bakeries in the
> rural areas using brick ovens use more
> local inputs, have lower costs of produc-
> tion, and lessen urban/rural inequalities,
> but at the same time it may be more diffi-
> cult to enforce (beneficial) social control
> (e.g., accurate bread weight) in small
> bakeries and they may exploit the labor
> force to an undesirable extent....Thus, a
> mix of techniques is probably socially ap-
> propriate, with local markets being supplied
> by small-scale bakeries....In evaluating the
> appropriateness of different types of baker-
> ies the possibilities for technological im-
> provements should not be ignored.[12]

That the selection of inappropriate technology
can have dire and far-reaching consequences can be
illustrated by recalling a historic event in 19th-
century India.

> In 1857, the native (sepoy) soldiers of
> Hindu and Moslem faith in the Bengal army of
> the East India Company were supplied with a
> new type of cartridge coated with grease.
> The latter was rumored to be the fat of cows

73

(sacred to Hindus) or pigs (anathema to
Moslems). The handling of these cartridges,
involving a literal biting of the bullet,
would have been sacrilegious under Hindu
and Moslem religious law. The Sepoy Rebel-
lion, for this and other reasons, broke out
in February of that year and could only be
crushed in March 1858, with great loss of
life on the British and Indian sides. Under
the circumstances, the choice of technology
had obviously been inappropriate, this time
for religious reasons.

To summarize, there is no single technology that
is appropriate for all purposes, and every technology
is suitable for reaching some objectives. When the
transferred technology does not "fit" into its new
surroundings--to the extent that this can be assessed
--it is because it may be unsuitable to the existing
technological base or economic resources available.
Or, it may be psychologically unsuitable by leading to
a fear of the unknown, of destabilizing change, and
thus increase resistance to it in a particular setting.
Last but not least, it may challenge the existing
value system or power structure.

Sociocultural Barriers
in Technology Transfer

There are several types of obstacles that can im-
pinge on the transfer of technology. Some are technological barriers preventing a potential recipient from
benefiting from a specific transfer at a particular
time. For instance, the absence of a transportation
network or other infrastructural facility may make it
difficult to import a steel mill which needs all these
facilities to be viable (or else they have to be
created from scratch). Others are legal barriers such
as restrictive contractual clauses imposed by the
technology supplier or regulations mandated by the
home or host governments (see Chapter 5). These legal
barriers can often be negotiated and a reasonable
settlement reached. Then there are barriers imposed
by the marketplace, such as prices beyond the immediate
reach of the would-be technology buyers, even though
funds can often be secured or suppliers' credit made
available to overcome this problem.

However, the most difficult obstacles to cope
with are probably sociocultural barriers, that is,
those having to do with the fact that different norms

74

and values may prevail on both sides of the transfer process. Generally, neither the technology supplier nor the recipient understand the other's sociocultural values and therefore the motivations of the parties involved in the transfer. This is especially so when the transfer has to do with different societies, say, modern versus traditional, and thus wide discrepancies in human values and in the norms that govern their behavior prevail. To one observer,[13] such differential perception of reality stems from the fact that

1. Different value systems generate differing concepts of right and wrong, proper and improper.

2. Different economic systems generate differing attitudes toward such concepts as competition, labor and capital efficiency, and acceptable standards of living.

3. Different societies make different assessments of the relative merits of job security and advancement.

4. Social and family customs may differently affect interpersonal relations and the individual's attitude toward group activities.

5. Personal relationships differently affect organizational patterns.

Such attitudinal differences as varying motivations and goals and different frustration levels can enhance the above difficulties by impinging on the risk acceptance level. There is also the natural resistance to change because of fear of the unknown, perceived threats to professional reputation, to economic or psychological security, or to existing organizational structures.

Communications can also present a problem since the most important method of technology transfer is through people, not documents or devices as was mentioned. On this score, more important than language difficulties are those involving thought processes since "the meaning of words is in us rather than in the word itself." Factor in that technology transfer involves individuals of different educational levels as well.

These sociocultural barriers often occur

simultaneously rather than singly and, again, can be synergistic.

Intranational Transfer

Even within the same society technology transfer may involve such roadblocks as communication barriers between inventors and innovators on one hand, implementers or entrepreneurs on the other. Here, too, there may be semantic difficulties even when everyone speaks the same language. Or, the sense or significance or urgency of the project may vary between transferer and transferee.

For instance, after the U.S. Air Force made jet technology (transferred earlier from Britain) available to the Boeing Corporation, the latter had to overcome Air Force reluctance regarding its usefulness in building jet tankers (as well as the opposition of commercial airlines afraid of being left with an obsolete propeller plane fleet) while it proceeded with designing the Boeing 707 passenger and KC-135 tanker plane (see Chapter 5).

Value differences between those inclined merely to add to human knowledge and those who stress the application, indeed, commercialization, of the project may also occur. Or, there may be a number of myths attendant on technology transfer having to do with cause and effect relationships--e.g., that labor-intensive technologies equate to increased employment opportunities. Too, there may be differences in assessing the technology transfer.[14]

An overview of public technology transferred from various U.S. federal agencies to state and local governments confirmed that among the variables related to overall success or failure were some of a sociocultural nature. For instance, the level of interdepartmental and project team communication; the frequency of government contact with customers/users; the degree of goal congruence; the amount of bureaucratic, political, and personal risk; the level of risk-averse behavior; the rigidity of state and local government practices mandated by civil service rules, line-by-line budget controls, centralized purchasing, or red tape; the degree of jurisdictional conflicts and suspicion of intergovernmental technical assistance; and the lack of training in science and technology of public decision-makers or the absence of an interdisciplinary sense among them.[15]

Accordingly, such features account for the success or failure of technology transfer in fields like police and fire protection, wastewater treatment, public transit, and resource recovery. Consider the successful transfer by NASA (the National Aeronautics and Space Administration) to local governments of a firefighting module, recycling asphalt process, emergency traffic routing; the U.S. Army's Edgewood Arsenal/ Natick Research and Development Center's transfer of new lightweight body armor; the U.S. Naval Weapons Center's transfer of an air pollutant analysis technique; the Department of Transportation's prototype buses; but also of social technology in such fields as law enforcement and education.

The study indicated that failure of the user to become involved in technology selection, the absence of personal interaction or of interaction between the user and the transferred technology, various forms of incompatibility, inadequate evaluation criteria relating to improved quality of service, greater citizen satisfaction, and at times lower cost in addition to technical and economic reasons explain failures in public technology transfer or why they do not take place.[16]

International Transfer

When transnational, that is, crosscultural, transfers are involved, these sociocultural barriers are usually magnified by such factors as the "not-invented-here" syndrome (referring to the natural suspicion of and resistance to anything foreign) and the totally different environment to which technology must be transferred. The following account by the former director of planning, Battelle Columbus Laboratories, will epitomize these difficulties.

> During the post-Korean war period I was extensively involved with the U.S. Army in the reconstruction program of the Seoul area. First, we were amazed to find that even though "we knew all of the answers," very few of them worked. Initially, we were simply insensitive to the (1) cultural differences, (2) indigenous motivating forces, and (3) different value systems.

> Often, one of the greatest mistakes Americans make...is the assumption that the response of foreigners can be predicted on

77

the basis of their own value system.

Often, one of the greatest mistakes technology transfer practitioners make is the assumption that a technology developed for one purpose and then offered for another will be viewed, used, and appreciated in a similar context.[17]

The need for sensitivity to sociocultural barriers in technology transfer will become evident from the following illustration, which also indicates how the overcoming of such barriers calls for adjusting the technology to the environment rather than vice versa, a theme that kept recurring in Chapter 3.

The Nuer are a central African tribe whom a British medical mission wished to vaccinate against smallpox. The Nuer wear no clothes, which is rather natural in their tropical climate. Their bodies display hash marks--scars which indicate their pedigree, rank (especially whether they are chiefs or merely followers), affiliation, and other "vital statistics." When the British appeared in the village, the local witch doctor who made the scars felt his vested interests threatened and caused the villagers to take to the bush. The British physicians had to strike a deal with the witch doctor so that the latter may add the vaccine when making the scars instead of the Britons using state-of-the-art airguns for the medical procedure which they had planned originally.[18]

If communication may be a problem within a society, logically this may be much more so between societies. Different standards of workmanship and varying specifications for commonly used items can prove to be major obstacles to successful technology transfer. Too, labor practices differ among countries, and skills taken for granted when production blueprints and specifications are prepared may be found in some completely different category or not at all elsewhere. Yet, all these matters--"our unique qualities, our experience, the language we speak, our nationality, our discipline"[19]--being part of the underlying culture of a nation, are seldom considered in producing documentation or designing products. For, being taken for granted, they may not intrude on the

78

consciousness of those involved on one or both sides of the transfer process.

To illustrate further sociocultural barriers in technology transfer, let us consider the Japanese-Indonesian joint venture to build a very large aluminum smelter and hydroelectric plant in Sumatra, Indonesia, in the 1970s:

> The Japanese were to provide highly exclusive technology and financing. Just as typically, they agreed to purchase the entire output on a long-term basis. The Indonesians were to supply the smelter and hydroelectric plant site as well as the local labor. The equity split was 90 percent for the Japanese and 10 percent for the Indonesians.

> When implementation got under way, the sociocultural problems between the two parties--both Asian--began to surface. In day-to-day matters there was little common ground. There was also a difference in the language spoken. Even though both knew some English, an interpreter-translator often had to be used. Their concept of time was also at odds with the Japanese being punctual while the Indonesians believed in "rubber time."

> So, too, were their approaches to decision-making. For while both the Japanese and the Indonesians sought consensus, the Japanese reached it in such a way that no one individual appeared to be responsible for managing the project while the Indonesians' consensus was much less spontaneous. Too, several Indonesians on the project were devout Moslems, and their ritual fasting on the job (presumably during the month of Ramadan) made them seem less than fully effective to the Japanese while the latter's perception of the Indonesians did not appreciate their religiosity.[20]

In other cases, the sensitivity of Japanese firms to sociocultural barriers in technology transfer is demonstrated by their use of Nisei employees (first-, second-, and third-generation Japanese immigrants) in Brazil to help bridge the language and cultural gap between Japanese techniques, their expatriate workers, and their Brazilian employees--to "facilitate the transfer of the Japanese management culture,"[21] as a Japanese executive explained. For the Japanese know that while specific technologies may be transferred on paper, their successful implementation normally resides in the minds of people. Thus, all other things being equal, there is less of a cultural gap to fill in the on-the-job training of Niseis. However, even here a Japanese equipment manufacturer noted that although the Niseis speak Japanese, they are Brazilian and their thought processes are already distinctly Brazilian.[22]

The Chinese solution of this problem is the use of abundant Chinese personnel in the field. Thus, while Western and Soviet technology transfer programs depend on local personnel to perform all but the professional jobs, the much larger input of Chinese expatriates living at a standard comparable to their Third World host country counterparts reduces friction between the transferers and recipients of the technology.[23] For instance, some 16,000 Chinese personnel were involved in the Tanzam Railroad project in East Africa, admittedly a sizable one where a 1,000-mile line was built at a cost of some $500 million.

An observer summarized these obstacles succinctly when he wrote:

Technology...is not universal. Technology is highly localized in that problems are defined in terms of interests, goals, and local culture of the organization in which they are being attacked. Similar technological problems may become defined in very dissimilar ways by organizations working on them because these organizations often have different objectives and value systems.[24]

FOOTNOTES

1. Mario Kamenetzky, "Choice and Design of Technologies for Investment Projects" (Washington, D.C.: The World Bank, April 4, 1982), p. 15 (mimeo).

2. Organization for Economic Cooperation and Development, "The Choice and Adaptation of Technology in Developing Countries" (Paris, O.E.C.D., November 1972), pp. 1 and 2 (mimeo).

3. Kamenetzky, "Choice and Design of Technologies,"

pp. 2-3.

4. Weiss, Jr., Mobilizing Technology, p. 2.

5. Rosemarie B. Thau and C. Wayne Bardin, "Fertilization, Pregnancy and Lactation," in Best and Taylor, eds., Physiological Basis of Medical Practice (Baltimore, Md.: Williams and Wilkins, 1984).

6. Jequier, Appropriate Technology, p. 20.

7. Kamenetzky, "Choice and Design of Technologies," p. 24.

8. Walter A. Chudson, The International Transfer of Commercial Technology to Developing Countries (New York: UNITAR Research Report No. 13, 1974), pp. 10-23.

9. American Council of Voluntary Agencies for Foreign Service, Evaluation Sourcebook (New York: ACVAFS, 1983), p. 4.

10. James Keddie and William H. Cleghorn in Christopher G. Baron, ed., Technology, Employment and Basic Needs in Food Processing in Developing Countries (Oxford: Pergamon Press, 1980), Ch. 3.

11. R. Kaplinsky, "Technological Choice in Bread Baking in Kenya," in Baron, ed., pp. 302-305.

12. Ibid., p. 304

13. Frank E. Cotton, Jr., "Some Problems Concerning Transfer of Technology and Management," State College, Mississippi, undated (mimeo), quoted in Charles H. Smith, III, Japanese Technology Transfer to Brazil (Ann Arbor, Mich.: University Microfilms International, 1981), pp. 23-24.

14. Glenn E. Schweitzer, "Technology Transfer and Development Mythology," in G. K. Manning, ed., Technology Transfer: Successes and Failures (San Francisco: San Francisco Press, 1974), pp. 138-144.

15. Samuel I. Doctors, ed., "State and Local Government Technology Transfer," Technology Transfer by State and Local Government (Cambridge, Mass.: Oelgeschlager, Gunn and Hain, 1981), pp. 23-24, 45 et seq.

16. Susan W. Woolston, "Local Government Technology

Transfer--Overview of Recent Research and Federal Programs," in Doctors, ed., p. 63.

17. Gabor Strasser, "Introduction," in Manning, ed.,
p. xxii.

18. David K. Evans, "Applied Anthropological Methodology as a Contribution to Technology Transfer Programs
within NATO," in Sherman Gee, ed., Technology Transfer
in Industrialized Countries (Alphen aan den Rijn, The
Netherlands: Sijthoff and Noordhoff International
Publishers BV, 1979), p. 354.

19. E. Bruce Peters, "Cultural and Language Obstacles
to Information Transfer in the Scientific and Technical Field," Management International Review (Wiesbaden,
Germany), vol. 15 (January 1975), p. 83.

20. Kathleen J. Murphy, "Third World Macroprojects in
the 1970s: Human Realities--Managerial Responses,"
Technology in Society, vol. 4, no. 2 (1982), pp. 138-
139.

21. Smith, Japanese Technology Transfer to Brazil,
p. 99.

22. Ibid., p. 102.

23. Janos Horvath, Chinese Technology Transfer to the
Third World (New York: Praeger Publishers, 1976),
p. 7.

24. Thomas J. Allen et al., "Technology Transfer as
a Function of Position in the Spectrum from Research
Through Development to Technical Services," Academy of
Management Journal (Mississippi State, Miss.), vol.
22, no. 4 (December 1979), p. 695.

CHAPTER 5

TECHNOLOGY TRANSFER AND PUBLIC POLICY

It was suggested earlier that technology transfer today is a major component of international trade. Now, according to classical economic theory, specifically the theory of free trade and comparative economic advantage, trading partners and by extension the world economy are all better off when they each specialize in what they do best and/or most inexpensively, exchanging those goods or services for others offered by those with different factor advantages.

However, economic rationality often conflicts with other desiderata so that public policy regarding foreign trade, specifically its technological components, is often modified by other factors, many of them containing a sociocultural or psychological dimension, some rational and others not. These factors condition the political decision-makers. Indeed, despite Adam Smith's arguments regarding the benefits of specialization and the division of labor, public policy decisions have never been shaped by purely economic considerations. For instance, in the 19th century the suggestions by British economist David Ricardo that his country should produce all steam engines, invented there, for the whole world went largely unheeded.

Today, the argument for technological self-reliance challenges an international division of labor in science and technology which leaves large areas of the developing world without an independent capability in this field, that is, without the means to create its own industrial base. In a number of instances, then, developing country governments have helped their public or private sector establish a more competitive indigenous industry in one or more fields and often berate their former or existing state of dependency which they claim to have been imposed on them through the exercise of monopolistic market power, especially by multinational corporations, on their territories, thereby inhibiting local research and development efforts and confining their participation mainly to cheap-labor intensive activities such as assembling instead of high-level technical functions.

These critics also point to the waste of scarce foreign currency and high prices mandated by the excessive importation of technology, and the inappropriate nature of many foreign products and services.

In short, such Third World leaders make a brief against the dominance maintained in trade and investment by technology exporters, mostly from the West.

However, even when these supposedly exploitative conditions do not prevail, technology transfer involves a number of trade-offs, of sacrifices, by all concerned. For instance, in the supplier firm or country, anticipated earnings must be offset against the creation of competition, the export of jobs, the possible loss of control over the transferred technology, and national security issues (discussed below). In the case of the user, a society's preference between the objectives of savings and economic growth on one hand versus more jobs, ecological or resource conservation, social justice, or cultural autonomy on the other will have to be weighed. Or, the trade-off may be between modernization as against other developmental goals. Now, since several of these trade-offs involve human, sociocultural, and psychological aspects, the governments have to consider these values, too, when they make public policy involving technology transfer.

Just whose values will prevail in any given case, whose views of appropriateness will apply, is usually the outcome of the force relationships or the power considerations at hand.

East-West Technology Transfer

In this essentially bipolar world, trade-offs are especially involved in East-West trade, the customary, generic way of referring to both West-to-East as well as East-to-West technology transfer, which increased spectacularly during detente.[1] Still, the latter fell to a fraction of its earlier total thereafter as the Reagan administration tried to plug "leaks" of Western technology, especially high technology, to the Soviet bloc, as the table below indicates.[2] A recent estimate places the loss in sales to the Soviet Union because of forgone trade to American companies at over $10 billion a year on account of government restrictions.[3]

Here, trade-offs are involved because questions of economic benefits must be offset against national security, foreign policy, human rights, or even considerations relating to terrorism. Accordingly, the positive consequences of technology transfer for some sectors have to be balanced against the negative

```
            Soviet Bloc's Foreign Suppliers

         Sources of Imported Capital Goods
            as a Percent of Total Imports

                              1975          1982

      East Europe

         Common Market        68.9%         62.2%
         Other Europe         21.7          26.0
         Japan                 5.7           8.3
         North America         4.3           3.4

      Soviet Union

         Common Market        57.6          38.6
         Other Europe         16.5          26.5
         Japan                14.7          32.2
         North America        11.2           2.6

   Source:  O.E.C.D.
```

consequences for others. This is especially true in East-West technology transfer because of the known ideological, economic, strategic, military, and political rivalry--in short, because of the power struggle --between the two global blocs.

In contrast to the extreme view, advanced as authoritative in the 1960s, to the effect that Western technology transfer has been the most important factor in the Soviet Union's economic growth,[4] subsequent research has considered the contribution of imported technology to its development as "small or uncertain."[5] Besides, the flow is in both directions and Soviet licenses have enabled the West to use techniques in mining and metallurgy, underground coal gasification, oil recovery, and particle acceleration.[6]

Furthermore, regardless of the technical or economic consequences of such technology transfer, America's Western allies also believe that East-West trade promotes peace by increasing prosperity and raising the cost of war. Naturally, such differences in views between the United States and its allies have led to a number of crises regarding which are the "critical technologies" that can contribute to the Soviet military effort, with Americans more likely to recall Soviet leader Nikita Khrushchev's quip that "buttons can hold up a soldier's trousers."

Among civilian transfers by the United States to the Soviet Union with possible military applications were the sale in 1972 of Bryant Centalign B precision grinders, which can be used to enhance the accuracy of missiles, and the Kama River truck plant (KamAZ) built mainly with American technology in the late 1970s. While this was by no means the first time that the United States was transferring automobile technology to the USSR--the initial American effort dates back to the late 1920s when the Ford Motor Company and others built the Gorkiy automobile plant[7]--the KamAZ project may have been the most significant.

The Kama River truck plant was built in Brezhnev, USSR, near the Urals industrial region. It was originally conceived as a sequel to the successful automobile plant built at Togliattigrad on the bank of the Volga River by Fiat of Italy and which now turns out over 700,000 Zhiguli, Lada, and Niva sedans a year. But why did the Soviets need the Kama plant? To change Soviet trucks from using gasoline to diesel oil (which is more economical) and to take pressure off the overloaded Soviet railroad system by increasing truck transportation.[8]

Besides the fact that neither Soviet roads nor service facilities were up to the proposed influx of heavy trucks, the political problems in readying the 40-square-mile factory were more formidable. For the Nixon administration in the early 1970s declined to let the Ford Motor Company or the Mack Truck Corporation act as general contractor for the Kama project, heeding the Pentagon's arguments that the factory could produce vehicles for military use. But even without Ford or Mack in charge, the Russians showed a preference for American technology by signing up the Swindell-Dressler division of Pullman to design and build the giant foundry, International Business Machines to provide a computer, while other contracts went to Westinghouse, Ingersoll Rand, and scores of concerns representing 250 contracts and $430 million (as well as to other Western firms for hundreds of millions more).

For in such fields as cybernetics, electronics, and military hardware, appropriate technology is also the most advanced technology since these fields are power-sensitive, that is, are closely linked to issues of national interest. Thus, as relations deteriorated following the Soviet invasion of Afghanistan, by 1980 supplier after supplier in the United States terminated the deal as the Pentagon charged that Kama trucks were being used in the operation.

In the case of such dual-purpose technology--civilian technology with possible military applications--the decision to transfer or not is especially difficult because it forces policy-makers to expose priorities they would often keep concealed.

Take the Dresser Industries case involving the proposed sale by the Texas-based multinational corporation of oil-drilling technology to the Soviet Union, a trade with a potential for both international technology transfer (between two countries) and horizontal technology transfer (from civilian to military application in the user country). The Soviet request to Dresser in March 1978 was that it help build a $144-million plant that would have increased Soviet drillbit production capacity by an estimated 10 percent and improve drillbit durability by 500 percent (the Soviets were already the largest producers of this item in the world, but their products were of inferior quality). Dresser's exports would have facilitated oil extraction by permitting deeper and more efficient drilling in such hostile environments as the Siberian permafrost where the huge Soviet oil deposits are located.

Now, in considering whether to approve licenses or not, the United States government looked at the transfer's strategic potential, end-use, and alternative foreign availability. Because the proposed Dresser sale would have involved tungsten-carbide technology (also suitable for armor-piercing projectiles) and an electron beam welder (for which several military applications were claimed by critics of the sale), the military implications of the deal were questioned. Too, whether it was in the national interest of the United States to help the USSR avoid a projected oil dilemma by, say, 1985, when the daily output of the world's largest oil producer is expected to decline to 10 million barrels of oil a day from the present 12 million but with a concurrent increase in consumption. The question was also raised whether, assuming little comparable foreign availability, oil-related technology exports should not be used as a political lever to pressure Moscow regarding human rights and other policy.

After lengthy controversy between those for and against such technology transfer, President Jimmy Carter changed his mind following the Soviet invasion of Afghanistan and in December 1980 revoked the approval he had initially granted in 1978. He had come to weigh the pros and cons in the trade-off differently than at the outset.

Two other cases are fairly analogous:

The Sperry-Univac case (involving a UNIVAC 1100/10 computer for TASS, the Soviet news agency) and the Cyril Bath Company case (involving stretch-forming presses for automobiles) also experienced difficulties and/or delays in obtaining licenses, the first because the computer has military potential, too, and the second because the stretch-form presses can be used to turn out aircraft parts and thus also fall into the restricted category.[9]

A somewhat similar scenario is in evidence in connection with the 2,800-mile Soviet gas pipeline which runs from the Siberian Urengoi field to several Western European countries and comprises 41 pumping stations along the way. The initial agreement involving this reportedly $10-

billion project had called for the Western
supplies of turbines and compressors to the
Soviets. But under the threat of American
trade sanctions on British, French, and
Italian companies that were helping the
Soviets to build the pipeline slated to
carry 40 billion cubic meters of gas a day
at its peak in 1986, many of the contracts
were canceled. The Soviets then had to do
with their own technology, untried in a
project of such mammoth size. There are
contradictory reports regarding its progress.[10]

Possibly like others, the Soviet Union has not
been above involvement in technology transfers "by
other means" (to quote Von Clausewitz). For instance,
after World War II, that country dismantled, shipped,
and reassembled entire German factories transferred as
war booty. The USSR has also been charged with in-
fringing on design and intellectual property, probably through reverse engineering, a synonym for technological plagiarism,[11] and it is bound to monitor
Western technical literature,[12] much more extensive
than anything comparable in the less open society of
the East. The Soviet Commission for Military Industry
specializes in such "transfers."

But, given the relatively few economic or political constraints in East-West transfers, even turnkey
plants, joint ventures, or licensed products can eventually compete directly in markets served by the original supplier firm. Thus, when the Italian automobile manufacturer Fiat built the Soviet Union's huge
Togliattigrad plant, it found that some of the Russian
Ladas were competing directly with cars produced at
Fiat's own facilities in Turin.[13]

Also, in the flap about the denial of restriction
of visas to students and academics from Soviet bloc
countries or the People's Republic of China suspected
of illegally trying to obtain technical or military
information, primarily on American campuses, objec-
tions have been voiced to the proposed ban on the
grounds that it would compromise the universities'
tradition of academic freedom.[14] Thus, even in this
marginal area there is an ethical dimension which must
be offset against a security aspect, that is, a trade-
off.

Further strictures--and with them other trans-

Atlantic rows--may occur when the U.S. Congress renews the Export Administration Act in 1984 enabling the federal government to impose export controls on economic, national security, and foreign policy grounds. The issue is confused because American businessmen, like America's allies, have been agitating for an easing in U.S. policy. Not only do some argue that trade is trade and therefore that technology transfer should be kept separate from political issues. But they also point to the fact that Western competitors often supply the technology to the East anyway in case of American default. For example, when the U.S. government cancelled the export licenses issued earlier to Armco relating to the building of a $353-million electric steel mill for the Russians, a French firm executed the order. Such conflict--ultimately based on differing values--is reflected within the U.S. government itself, with the Department of Defense generally pressing for a strict and the Department of Commerce for a liberal trade outlook.

Since debates of this sort about the desirability of technology transfer often defy quantitative analysis or proof, it is easier for ideological or sociocultural or political considerations to overshadow factual ones. Public policy will then have a high human values content.

West-West Technology Transfer

The issue of trade-offs arises even between friendly countries since competition has increased on the international marketplace and even importers of military technology are more sensitive to the economic and other fallouts of large orders of noncommercial hardware and software which, in the aggregate, may well exceed total commercial technology transfer.

For instance, the buyers of military technology wish to use coproduction for broader development objectives--say, to train skilled manpower. Now, while coproduction allows for division of labor and specialization between two or more partners, the license-exporting country may also have to share some of its know-how and lose part of its technological advantage. Too, the chances of technology falling into unfriendly hands are increased (recall the case of the Chinese pilot who defected with his Soviet-built MiG plane to Taiwan), mandating the need for even larger investments in military research and development. Witness, for example:

Spain's decision to purchase the American F-18A Hornet combat aircraft. The American producer, McDonnell Douglas, had to compete with the challenge of France's Mirage 2000 and the European jointly-built Tornado. One of the crucial ingredients in this "deal of the century" for Spain was the transfer of advanced aeronautic and electronic technology from the American manufacturer. For a while there was reluctance on the part of McDonnell Douglas, and the Spanish press complained that "our offsets may end up consisting merely in our sale of shoes, oranges, and tourism."[15] Eventually, however, an agreement was reached and in May 1983 the Spanish Air Force decided to purchase 72 F-18A fighter-bombers after the American firm agreed to have some of the plane's parts produced in Spain under license. The point is that, under increasing Western European competition, American technology suppliers may have to abandon their preferred form of technology transfer (export sales in this case) and allow for some joint production of components (under license). A similar agreement was worked out for the sale of 12 Harrier AV-8B jump jets to the Spanish Navy produced jointly by McDonnell Douglas and British Aerospace.

Or take the case of Japan's aircraft industry:

The Japanese are learning now to make good civilian jet aircraft for the international market despite the fact that after World War II they lost seven years of aerospace development when they were forbidden to have an aeronautical industry. Thus, some major components of the Boeing 767 jetliner are already being produced by Kawasaki Heavy Industries and Mitsubishi Heavy Industries of Japan after the latter shared in the development costs of the front and middle sections of the airframe and wing. The Boeing Company and the Japanese are also negotiating a partnership for the development of Boeing's new-generation 150-seat fuel-efficient passenger aircraft, known in Japan as the YXX, even though Boeing is concerned that the Japanese may use their acquired expertise to eventually control the market.[16]

91

This is true in other fields, too. For instance:

In missiles some observers believe that the
ASM-1 (air-to-surface missile) developed by
a group led by Mitsubishi Heavy Industries
is a superior product. Japanese-developed
electronic miniaturization is beginning to
make its mark in such applications as portable missiles. Then, the Japanese have developed the detection absorbers that allow
the production of stealth aircraft. The
Japanese are also working on a tank, to be
completed in the late 1980s, expected to be
second to none.

Given Japan's extensive science and technology base
with its research and development capability, the
transfer of technology to Japan seems to be more of a
convenience to the latter, saving time and money, than
an absolute necessity. After all, had Japan not built
one of the best fighter-bombers in the 1930s making
its attack on Pearl Harbor, Hawaii, possible in 1941?

Albeit, whether for security or economic but especially political reasons, the Japanese, too, have
decided to allow the export of some of their military
technology to the United States. Involved are robots,
laser, very large-scale integrated circuits and other
components for weapons, fiber optics for communications, next-generation computers, and many high-
quality conventional products. Even though such military technology transfer is aimed to ease American
protectionist measures, the Japanese Cabinet noted
that "'with the recent advance of technology in Japan,
it has become extremely important for Japan to recip-
rocate in the exchange of defense-related technologies
in order to insure the effective operation of the
Japan-U.S. security arrangements.'"[17]

But such a resolve has been accompanied by Japanese fears, re-echoing American ones, that the American
intention is to undercut Japan's competitive edge over
the United States in certain high-tech fields under
the guise of arrangements on military know-how transfers. There is also apprehension that the United
States wants to "fit" Japan into NATO's "family
weapons system" so that Japan would end up as a sub-
contractor supplying parts to the United States and
buying completed weapons in return. So the question
remains whether any of this technology will ever
emerge for transfer from the muddy waters of politics.[18]

Beyond economic or military motivations, governments may promote arms transfers for political purposes, that is, to meet national foreign policy goals
by maintaining some control over the behavior of
other countries. Thus, arms may be sold as a reward
for the recipient's friendly policy (e.g., the United
States selling equipment to Saudi Arabia or Israel),
to create a degree of dependence on the supplier
(e.g., the Soviet Union's sales to Iraq), to maintain
a balance of power in a given area (e.g., the United
States' transfers to Egypt and Israel), or so that the
recipient may perform surrogate functions (e.g., the
United States' transfer to what used to be South Viet-
nam, the Soviet Union's transfers to Syria and Cuba),
and for other reasons. The political risk involved is
that a change in regime may make the former supplier
the new enemy (e.g., the United States after Somoza's
Nicaragua fell to the Sandinistas or Iran to Khomeini)
and that the transferred equipment may end up in the
hostile hands of second or third parties. In all
cases, more than merely technical or economic considerations are involved.

North-South Technology Transfer

The doctrine of mercantilism that prompted the
mother-country to turn raw materials from the colonies
into manufactured products which it then sold back to
these kept the latter--representing the greater part
of the international political map till after World
War II--in a state of dependency. It also explains
the existing lack in these recently independent countries of a scientific and technical base and thus of
industrialization. For even when manufactures did
exist, Western interests if not competition put an end
to them--say, the Indian textile mills which had
thrived prior to Britain's colonization of the subcon-
tinent in the 18th century. Admittedly, with the low
purchasing power of the masses and thus the weak domestic markets, lack of savings and thus of capital
investments, these dependencies were hardly promising
candidates for industrialization and economic growth.

Accordingly, the transfer of capital from the
North to the South--roughly, from Western, developed
countries to developing countries primarily in Asia,
Africa, and Latin America--gave way to bilateral or
multilateral technical assistance and finally full-
blown commercial technology transfer, aided and
abetted by Third World governments which provided a
whole range of tax, tariff, depreciation, and other

incentives to speed the process. Basically, the strategy of the Third World, loosely categorized as the South, was at first to seek material and human resources for import-substituting industries, initially assembling these products locally and then hoping to produce the components and finally even the plant and equipment to turn them out. But regardless of the description--whether aid or technical assistance or technology transfer--the use of imported equipment often involved expatriate personnel in the recipient countries where maintenance and other skills are characteristically low and where abuse or misuse of equipment leads to its disuse and therefore to the failure of the particular project.

Too, North-South technology transfer, whose volume is far less than West-West (or North-North) flows, was in some cases inappropriate both in terms of the methods of production (capital- and skill-intensive, labor-saving, large-scale) and the nature of products involved aimed at higher-income consumers, much of it very expensive and of relatively little social value. Thus, it is not surprising that the result of such technology transfer provided less than a proportionate increase in overall employment compared to output; led to a brain drain given the high level of technological unemployment and underemployment in the South; the training of personnel whether abroad or at home in skills that were often irrelevant to these developing countries outside of the supplier multinational corporation's subsidiary; a lack of real wage increases and thus a growing maldistribution of income; increasing differentiation among social classes instead of their greater integration, often despite acceptable overall growth rates. That is, the changes in social structures with which development (as opposed to the narrower concept of economic growth) is associated frequently did not occur. And indeed, the technology recipient's environment contributed to this lack of obvious results by unsuitable research and development rates and conditions; the lack of scientific outlook by scientists, technicians, and engineers; and nonindustrial attitudes by workers.

It is in the wake of these indifferent results and the recognition that political independence in the post-World War II period did not bring economic autonomy and even less technological independence but rather, since consumption patterns and production methods originated from technology suppliers, cultural and psychological dependence as well that many Third

World governments entered the fray in a bigger way. They did so to improve bargaining since the firms involved (especially when the buyers are closely identified with the sellers as is undoubtedly the case when the same multinational corporation is both supplier and recipient) have little knowledge about negotiating an appropriate agreement.

The problem of how these countries which suffer from the vicious circle of poverty--low savings, low investments, low production, low income--in a setting of high birth rates can be shunted into a virtuous circle of economic growth based on science, technology, and thus improved productivity (input/output ratios) is still looking for a solution. And experimentation with self-reliance strategies, often linked to the use of appropriate or intermediate technology and the unbundling of technology transfer providing access to the elements underlying the package of scientific knowledge, engineering, marketing, and other components of technology transfer, is still inconclusive. However, some Third World countries--South Korea, Singapore, Taiwan, Hong Kong, Brazil, and Mexico in particular--have achieved noteworthy economic growth rates.

In recent years, what with the depressed economic setting in developed countries themselves and the indifferent effect of imported technology on developing local capabilities for creating new technology, the South has tended to shift away from looking to the North for technology and is now seeking more to extend its own endogenous capabilities than trying to control powerful foreign technology suppliers.[19] But it must be recognized that the policy of both technology sellers and buyers is generally based on value-related judgments--for example, that high profits and political/military power should be maximized for the former and that self-reliance and the acquisition of appropriate technology (however defined) are desirable for the latter. This capsule of the situation will now be broken down among its economic, sociocultural, and political components.

Economic Components of Public Policy

Despite a few success stories such as those mentioned, the gap between the North and the South has been widening as measured by such standard gauges as per capita income. Still, there have been improvements in developing countries--necessarily so since technology transfer has enabled a number of them to do things better than beforehand. More importantly, technology transfer has sensitized these underprivileged countries to the possibilities of industrialization through their exposures to state-of-the-art techniques which, for better or worse, enable them to cope more effectively with the demands of the modern age. Admittedly, this assessment would have to be discounted by the effects of sizable arms transfers from North to South, which are often of questionable economic or social value. Indeed, the multiplier effect of civilian goods in the Gross National Product is higher than of military hardware or software, quite apart from obvious noneconomic consequences.

Still, the fact that some newly industrialized countries (NICs) have "made it" seems to suggest logically that others could follow. But this can only be if developing countries concentrate on making the educational and research-and-development systems appropriate to sustain integrated and continuous growth in technological capacity and on creating the scientific and sociocultural context in which industry can flourish.

For instance, there is an excess number of university students in developing countries taking law and liberal arts curricula instead of science or engineering. Too, the entrepreneurial spirit is often discouraged by a political preference for nationalized industries or the paralyzing effect of red tape with which business has to cope. And as mentioned, at the microeconomic level technology transfer often increases the earnings gap between the few affluent members of developing societies and the typically larger number of poor people.

Sociocultural Components of Public Policy

North-South technology transfer has had even more far-reaching sociocultural than economic consequences. By creating industries in cities, multinationals, highly efficient technology transfer agents, have often disrupted the existing social system and contributed to the proletarianization of the peasantry or merely attracted rural folk to urban centers in search of employment. These then become wage earners (in case they find jobs) but also suffer from the alienation of an unfamiliar and brutalizing industrial environment. Or, they join the ranks of an uprooted, unemployed labor force, especially as extractive

industries have become more capital-intensive in recent years. The fact that technology transfer agents
tend to pay higher wages, especially as they use a
larger percentage of skilled or semi-skilled manpower
than local firms, may create a labor aristocracy and
thus greater class differentiation. The Islamic fun-
damentalist revolution in Iran epitomizes a revulsion
against many of these developments attendant on hasty
modernization or at least industrialization and urbanization.

Now, while an elite of some sort may benefit dis-
proportionately from technology transfer in terms of
financial returns, conspicuous consumption, and in
other respects, so have some Third World ethnic groups
compared to others.[20] Not surprisingly, studies suggest that Japanese multinationals have shown a marked
preference for hiring Japanese-Americans in Hawaii,
and there are analogous situations involving British,
Chinese, and other expatriates. Such preference would
translate itself into higher earnings by these ethnic
groups with whatever political stabilizing or destabilizing effects this might entail. Or, the policies of
the technology transfer agent may have an impact on,
say, the ethnic, racial, or religious policies of the
host government. For instance, under pressure from
American multinational corporations, the South African
government has made some concessions regarding the
application of apartheid laws to black workers. Not
so the Saudi government when it comes to the use of
Jewish employees by multinationals in that traditional
Arab country.

Or, technology transfer may benefit specifically
some locality or region of the recipeient's country--
say, where a facility is installed. This has created
enclaves of prosperity amidst poverty (shades of Aca-
pulco!), accentuated the disparity between urban and
rural areas, heightened differences in consumption
patterns and lifestyles between the affluent and the
underprivileged, or altered the tastes of the latter
to less appropriate ones. The story is told of how
the masses in Jamaica changed their breakfast consumption habits from fish and bananas (plentiful and
cheap in this Caribbean island) to Kellogg's breakfast
cereals while large quantities of bananas spoiled for
lack of markets.[21]

Less obviously, technology transfer may have a
gender bias built into it, especially in the characteristically traditional societies of developing

countries where sex roles are more highly differen-
tiated than in modern societies. For instance, taxi
drivers in the Third World are rarely female because
sociocultural restraints prevent them from taking advantage of this form of livelihood consequent on the
transfer of automotive technology there.

Political Components of Public Policy

Technology was not viewed as a major policy issue
till the early 1960s when, at the first United Nations
Conference on Trade and Development (UNCTAD I) in
1964, technology transfer was discussed in a general,
albeit limited, context. By UNCTAD II in 1968, however, most of the themes of the technology transfer
debate had emerged: The need to adapt imported technology to the conditions of developing countries; the
desirability of developing countries' generating their
own technological capacity; the adverse effects of
technology transfer on their balance of payments; the
potentially disruptive role of multinational corporations; revision of the international patent system;
and the elimination of restrictive business practices.
Subsequently, further technology-related issues were
added in various forums.[22]

The upshot of the fact that technology transfer
has now been on the international political agenda for
over two decades has been the increasing politiciza-
tion of technology-related issues. For instance,
charges have been made by some developing country
leaders that intermediate or soft or light-capital or
progressive or appropriate technology is tantamount to
second-rate technology and that the West's promotion
of same is a ploy to keep Third World users dependent
on the First World. Or that the forms of North-South
technology transfer, with emphasis on turnkey operations, help maintain the monopoly of the former.

Yet, some radical opinions posit that since what
is important is the amount of goods produced rather
than the number of jobs created, it is the former and
not the latter which establishes the level of social
welfare as well as economic and political independence, the most advanced technology which multinationals have is the best. In this view, an inappropriate
technology would be an underdeveloped technology which
is seen as underlying economic underdevelopment and
dependence. Hence, it is the transfer of state-of-
the-art technology which accelerates the development
of Third World countries and which should be pushed.[23]

While there was a time when negotiations on the drafting of the Code of Conduct for Transnational Corporations and that of the International Code of Conduct on the Transfer of Technology were hot issues in the North-South dialogue,[24] this is no longer as true because of the reasons mentioned earlier. However, the issue of devising new international mechanisms for altering the existing patterns of technological trade,[25] though greatly muted, is not quite dead. For the South still reasons that it has such restructuring, including a slew of preferences, among them in technology transfer, coming by virtue of the past imperialist exploitation and injustice perpetrated against it when its members used to be colonies, and by the fact that their structures have been fashioned by the international division of labor, closely controlled by the technologically advanced economies.

Thus, the New International Economic Order (NIEO) under which all these issues (and a few others unrelated to technology) are subsumed is slated to help the newly independent and underdeveloped Third World catch up through several concessions by the North, technology being now viewed by the South as part of the universal human heritage to which all countries deserve access. Even so, the point has been made that given the heterogeneity of the North but especially the South, the North-South battle line to discuss these issues is not relevant as it leads to the concealment of many issues and a state of ambivalence.[26]

But in the meantime, the North has resisted many of these demands of the South. This is partly because the former generally fails to recognize the injustice caused to the South by the system and, more practically, the North has its own problems of continuing high unemployment, inflation, increased competition (including that from the South, better able to produce more sophisticated items), and others. Occasionally, the North explains this resistance to the NIEO[27] by stating that economic growth is primarily an internal matter dependent on domestic savings and investments as well as population control and so on rather than international trade (including technology transfer). In this view, any forced sharing of wealth by the few rich countries with the many poor countries would only yield negligible per capita gains.[28]

Thus, while disagreements will continue on these

99

and other matters--where the emphasis should be in development goals (agriculture versus industry) or which
system is the most conducive to development (centralized planning and nationalized industries versus a
market system and privately-owned firms)--technology
transfer is certain to remain on the bilateral, regional, and multilateral political agenda. Only the
form of these future discussions is in question,
namely, whether developing country officials will keep
on arguing the moral liability of developed countries
versus their own entitlements or focus on the solution
of underlying technical, economic, human, and sociocultural problems.

East-South Technology Transfer

That the problems attendant on North-South technology transfer are not far different from those involved in East-South trade becomes evident in this example of the Bokaro steel plant, the second to be
built in India with Soviet aid in the 1950s and 1960s.

The contours of this gigantic project,
located some 250 miles from Calcutta,
changed from a plan drawn up by the Indian
Planning Commission calling for only minimal imports of equipment and services of
steel technologists, expected to be financed
by American aid, and a major role for an Indian private sector consulting firm, N.M.
Dastur and Co. (Dasturco), even though the
steel plant was to be a public project, to
no U.S. aid because of the American government's reluctance to encourage the Indian
public sector and its socialist goals and
the opposition of American steel and shipping interests fearful of increased competition. Rather, as in the case of the Aswan
High Dam in Egypt, India reached an eventual
agreement with the Soviet Union for aid-
financed equipment and large-scale technical
assistance. The Soviets managed to muscle
out Dasturco from a role in the final aid
agreement after negating most of its recommendations for changes made by the private
Indian consultants relating to the Soviets'
engineering plan on the grounds of design
and equipment, need, and cost estimates.
The Soviets justified their rejection of
most of these recommendations, indicating
that the advantages claimed for them were

in fact marginal or nonexistent or that
they were based on inaccurate estimates or
that cost savings resulting from their
adoption would be nominal. For instance,
Dasturco's recommended savings of 1,075
million rupees were pared down to 95 million rupees by the Soviets.

The Indian policy-makers gave in to the
Soviets' insistence on building a larger
plant than the Indians felt they needed (be-
lieving that prestige flows from size) for
the sake of expanding a basic industry in
the public sector (which enhanced the rapport between Soviets and Indians), the potential of import substitution (also pleas-
ing to the Russians since the Soviets had
built a number of the ancillary facilities
which the latter wished to see efficiently
utilized) and the saving of scarce foreign
currency. The Indians also agreed to trade
off state-of-the-art technology in steel
smelting and slabbing (with which the Soviets were not comfortable) and the use of
local designing, engineering, and managerial talent for the sake of obtaining what
they viewed as a last-resort foreign aid
from the Soviets.

Because of this, the Indians saw their
bargaining power reduced vis-a-vis the Russians, who now linked their supplies of
equipment to their exclusive consulting services, technical assistance, and complete
control over the project.

But by insisting on such a predominant
role for themselves with the corresponding
squeezing out of Indian talent and expertise, even of the use of local equipment and
material (e.g., refractory ores), the Soviets undermined India's goal of self-reliance
and indigenous industrial development. Too,
the Soviets' demands for sumptuous pay and
allowances, overly comfortable living and
working conditions, and fringes for the
larger number of Soviet personnel contrac-
tually mandated on the Indians had the effect of cooling relations between the two
partners for a while.[29]

What can be deduced from this case is not whether technology transfer from one source rather than another is free from attached strings or not but rather what and whose sociocultural bias is at play and whether and how it differs from someone else's. It would be interesting to speculate which value-related biases would have been built into the Bokaro project if the technology and financial aid had been supplied by the Americans instead of the Soviets, as originally contemplated. For the policy involving technology transfer is rarely value-neutral. Only its shape and nature change from agent to agent, from case to case, from time to time.

Public Policy Problems in
Technology Transfer

International technology transfer has been seen as raising the following political issues:

1. It exports jobs through licensing or direct foreign investment;

2. It accelerates foreign competition in the world market;

3. Technology suppliers are setting too low or too high a price (depending on perspectives) for their products or services;

4. Technology transfer will ultimately result in an international division of labor which would narrow the national economic base with resulting serious implications for economic and military security; and

5. Civilian (especially dual-purpose) technology transfer will be used for military purposes abroad.[30]

Now, governments act on the premise that market forces alone cannot be relied upon completely to bring about optimal conditions of technological development or technology transfer along socially desirable paths. For while market forces have to do primarily with the adoption of microtechnology whose spread is unplanned and undirected but occurring on the basis of ad hoc decisions by large numbers of consumers, microtechnological projects can still have very significant sociocultural consequences (see Chapter 2). Hence, governments frequently regulate both sides of the transfer process by approving or disapproving agreements even where private parties are involved or by promoting transfer agreements directly.[31]

Even at the domestic level, the role of government has become eminent either as a large potential customer itself (thus providing reasonable assurance of an adequate market for products) or by granting patent protection encouraging the transferer to enter the commercial market. For instance, in the United States, some major examples of successful commercial transfers tabulated below all benefited from various degrees of such positive federal government involvement:[32]

Table 5.1

Technology Transfer Cases
Promoted by the U.S. Government

Type	Period	Transfer Originator/ Recipient	Technology Transferred
Corporation to corporation	1952-54	Bell Laboratories/ Texas Instruments	Transistors
Government to corporation	1952-59	U.S. Air Force/ Boeing Corporation	Jet aircraft
Technical institute to corporation	1947-53	Battelle Memorial Institute/Haloid (later Xerox) Corporation	Xerography
University to corporation	1946-51	University of Pennsylvania/Univac	Electronic digital computers

But government's role in technology transfer can also be negative. Consider, for instance, the case of Coca Cola being banned in India because the Coca Cola Company would not disclose the secret of its famous syrup formula even against compensation, or the threat of nationalization in Chile during Salvador Allende's regime.

Historically, the single most important impact of technology on the international political system has involved the distribution of power among its units. As an element of power flowing from a country's industrial base and thus impacting on its military capabilities, the possession of technology has led to control and outright domination. Indeed, the outcome of several wars--e.g., the defeat of Japan in 1945--can be traced to the successful use of more advanced technology by the victor (atomic bombs in this case). And while military power has become less relevant in this age of insurgency wars and "cheap," terrorist wars, technology transfer is still a reflection and reinforcement of the existing political power structure and dominant political interests. This is so even though the advent of nuclear weaponry may have changed war as an instrument of foreign policy "by other means" from a zero-sum situation (where one side's gain is necessarily the other side's loss and vice versa) to a non-zero-sum game (where both sides can come out ahead), primarily because in this field the optimization of security calls for cooperation rather than rivalry, making political resolve more important than technology.[33]

The role of government can be perceived in Baranson's analytical model in Figure 5.1. This model suggests that in an international transfer of technology, the motivations of four parties must be analyzed. That of the technology supplier, that of the recipient, and those of their respective, home and host, governments. Each has different objectives, and an effective transfer can result only from a mutual, even a common, understanding.

The motivations of a technology supplier firm have already been detailed--growth and profit maximization through the securing of reliable sources of raw materials and market outlets and lower-cost production facilities, the avoidance of various types of barriers such as import restrictions, and so on.

The motivations of the technology recipient have also been discussed--faster availability of technology than if it were developed locally, a better competitive position in local or third-country markets, standard of living or developmental considerations (in the case of public agencies or international organizations), and so on.

The home government of the supplier will

FIGURE 5.1

Supplier Enterprise Government		Technology		Purchaser Enterprise Government	

● Government
  Policies
  – Economic
  – Political
  – Strategic

● Bargaining
  Power
  – Government controls
    over enterprise ac-
    tion
  – Technological lead
    of enterprise

● Enterprise
  Strategies
  – Shift from equity in-
    vestment and manage-
    ment control to sale
    of technology and
    management services
  – Measured release of
    core technology
  – Release of technology
    no longer central to
    company business
  – Necessity to accept
    foreign affiliate due
    to enormity of R&D or
    capital investment costs,
    offset requirements,
    scale of operation re-
    quires consortium

● Quantum and
  complexity

● License to manu-
  facture or turn-
  key plus

● Operative-
  duplicative-
  innovative

● General-firm-
  system specific

● Stage in product/
  process cycle

● Government
  Policies
  – Economic
  – Political
  – Strategic

● Bargaining
  Power
  – Absorptive capa-
    bilities
  – Alternative sources
    of technology
  – Astuteness
  – Financial resources

● Enterprise
  Strategies
  – Internationally
    competitive tech-
    nology
  – Duplicative and/or
    innovative design
    engineering capa-
    bilities
  – Market entry to
    export exports
  – Training of tech-
    nical managerial
    manpower
  – Fast, efficient
    technology trans-
    plants

SOURCE: Jack Baranson, Technology and the Multinationals: Corporate Strategies in a Changing World Economy (Lexington, Mass.: Lexington Books, D. C. Heath and Company, 1978), p. 14.

generally view the transfer for its effect on the
local economy, including the employment situation. Or,
conversely, it may wish to restrict the exports of
specific technology for economic, strategic, or political reasons.

Finally, the host government will consider the
effect of the transfer on its balance of trade and/or
payments accounts or its foreign exchange reserves as
well as for developmental, strategic, or political
reasons. Like the home government, the host government may also wish to restrict the transfer of technology for its own reasons.

In effect, this analytical model infers that
there is a close linkage between a given technology
and a specific political economy so that technology
and technology transfers are a reflection of and reinforcement to existing dominant political interests.[34]
This is so because, associated with each type of technology, there is a particular distribution of benefits, of rewards, for the technology suppliers and for
its users. In practice, at least, there is a close
connection between a chosen technology and its distri-
butional consequences.

Because of this, the distribution of benefits
flowing from the use of advanced technology is completely different from the distribution of benefits
from alternative technologies such as appropriate
technology. In the case of advanced technology, the
beneficiaries are a small elite connected with the
producers of such technology or who use their products
or services. In the case of alternative technologies,
the beneficiaries are small-scale local producers, the
potentially unemployed or underemployed, and low-
income consumers who provide a market for low-priced
appropriate technology products. Therefore, the beneficiaries from advanced technology (usually suppliers
in developed countries and the elite in the recipient
countries) generally do not benefit from appropriate
technology transfers and the widespread use of appropriate technology would involve a major shift of resources from the elite-dominated modern sector to the
popularly controlled traditional sector. Considering
the fact that most Third World government structures
are not pluralistic but rather elitist, it is not surprising that the advocates of advanced technology are
better represented in their policy-making organs than
the champions of alternative technologies.

106

Technology Transfer Regulation
and Human Values

Public policy-makers, bent on technology-for-
economic-growth goals or technology-for-consumption
goals, are sometimes slow to become sensitive enough
to human or environmental needs. This can happen when
efforts in business and/or government are confined to
promoting evolution within a given technological field
without reference to others. The result is a closed-
circuit orientation which does not necessarily represent the public's interest. This can be modified if
the goals of technical development and transfer are
defined with increasing public participation and reflect the latter's order of priorities, of values. And
as technology development changes--and it often does
under government impetus--so does the content of technology transfer which, it was stressed, has to do with
the evolution of people and societies resulting in a
variety of interdependent changes--and disequilibria.
These interdependencies between political systems and
their values, between the nature of technology and its
transfer process, are complex.

Too, technology transfer cannot be a neutral process since it affects the distribution of power within
societies. Accordingly, it is used, more or less explicitly, for political purposes by the various participants in the social systems.

It was seen that technology transfer is much more
politicized when the actors are divided by ideology
(East-West transfers), international competition (West-
West transfers), but also different levels of development (North-South and East-South transfers). Indeed,
a focal point of the problem is: How can a beneficiary acquire the technology without being dependent on
rules of the game set by the suppliers? This must
necessarily remain a rhetorical question. What is
stressed here is that domestic government policies
condition the consequences of a particular technology
transfer. For domestic policies to be successful,
they must minimally fit into a society's sociocultural
and psychological makeup. As for international regulations, they transform technology transfer into a
privileged mechanism used by the more powerful to control their relations with the less powerful.

It was also seen that technology transfer in
strategic areas (e.g., defense) is much more closely
regulated than in the commercial sector where the

107

market forces of demand and supply more frequently
prevail. Albeit, when governments restrict the ability of foreign firms to produce on their territory,
there is a greater likelihood for cross-licensing or
joint ventures to get around the difficulty. For instance, in the 1960s, the major Japanese computer
companies--Hitachi, Mitsubishi, Nippon Electric, Oki,
and Toshiba--entered into licensing agreements with
RCA, TRW, Honeywell, Univac, and General Electric,
respectively, because of such restrictions.

Brazil went at least as far as Japan in this respect. There the government regulated imports and restricted direct manufacturing by multinational corporations in order to earmark the most dynamic sectors of
the computer industry for local production, specifically, micro and minicomputers. Furthermore, Brazilian
law bans most types of restrictive clauses in licensing agreements. The government has also provided protection for local technology development by turning
down projects using foreign technology whenever local
design capabilities for microcomputers, low-speed
modems, and bank terminals were available.[35]

Consider, too, how a Taiwan-based electronics
manufacturer, Gamma Corporation, established a television set plant in Singapore to circumvent the quota of
the European Economic Community and also to supply
completely knocked-down units to high-tariff countries
in Southeast Asia.[36]

While government policy normally strives to provide incentives for technology exporters or incentive/
protection for importers, it was noted that inconsist-
ent regulations may act as a damper on technology
transfer and thus on development. This seems to have
been the case in India.[37]

Summation

Public policy is not neutral but is socioculturally based in that it subsumes a number of factors--
especially value-related ones--which encourage or discourage, promote or ban, technology transfer.

In the last analysis, public policy vis-a-vis
technology transfer mirrors the balance of power, internal and external, of decision-makers and can be
used in turn to reinforce such power. For this reason, the symbiotic relationship between technology
transfer and human values is paralleled by the

interaction between technology transfer and public policy.

1. Marilyn L. Liebrenz, Transfer of Technology: U.S. Multinationals and Eastern Europe (New York: Praeger Publishers, 1982), p. ix.

2. Reproduced from The New York Times, May 20, 1984, p. E3.

3. Raymond Bonner, "U.S.-Soviet Trade Bars Said to Cost $10 Billion," The New York Times, May 25, 1984, p. D1.

4. Especially the seminal work of A. C. Sutton, Western Technology and Soviet Economic Development, vols. 1-3; 1917-1965 (Stanford, Calif.: Hoover Institution, 1968-73).

5. Paul Lewis, "The Costs of Selling U.S. Technology," The New York Times, May 20, 1984, p. E3, quoting a 1984 O.E.C.D. study.

6. R. J. Carrick, East-West Technology Transfer in Perspective (Berkeley, Calif: Institute of International Studies, University of California, 1978), pp. 56-57; V. Sobeslavsky and P. Beazley, The Transfer of Technology to Socialist Countries: The Case of the Soviet Chemical Industry (Westmead, England: Gower, 1980), p. 109; Eric Hayden, Technology Transfer to Eastern Europe: U.S. Corporate Experience (New York: Praeger, 1976).

7. George D. Holliday, Technology Transfer to the USSR, 1928-1937 and 1966-1975: The Role of Western Technology in Soviet Economic Development (Boulder, Colo.: Westview Press, 1979), Ch. 5.

8. Serge Schememann, "Brezhnev Souvenir: Vast, Limping Truck Factory," The New York Times, February 4, 1983, p. A2.

9. All three cases are reviewed in Gary K. Bertsch, et al., "Decision Dynamics of Technology Transfer to the U.S.S.R.," Technology in Society, vol. 3, no. 4 (1981), pp. 412-415. In a more general way, the problems of Western technology transfer to Eastern European countries are covered, inter alia, in U.S.

Congress, Joint Economic Committee, Issues in East-West Commercial Relations: A Compendium of Papers (Washington, D.C.: U.S. Government Printing Office, 1979), especially pp. 1-124; also in Eugene Zaleski and Helgard Wienert, Technology Transfer Between East and West (Paris: Organization for Economic Cooperation and Development, 1980).

10. John F. Burns, "Is the Soviet Pipeline Completed?" The New York Times, January 5, 1984, p. D1; John F. Burns, "Soviet Confirms Fire at Gas Pipeline," The New York Times, January 12, 1984, pp. D1 and D2.

11. Raymond S. Mathieson, Japan's Role in Soviet Economic Growth: Transfer of Technology Since 1965 (New York: Praeger, 1979), p. 234.

12. Carrick, East-West Technology Transfer, pp. 16, 60, and 72.

13. Gee, Technology Transfer, p. 33.

14. Kim McDonald, "U.S. to Restrict Visas for Visitors Likely to Obtain Data Illegally," The Chronicle of Higher Education, vol. 26, no. 12 (May 18, 1983), p. 21.

15. J. L. Jurado Centurion, "Future Combat and Attack Aircraft: Economic Offsets, the Key in the Negotiations," El Alcazar (Madrid), December 30, 1982, pp. 10-11 (in Spanish).

16. Pamela G. Hollie, "Boeing's Touchy Japanese Tie," The New York Times, January 29, 1983, pp. 29 and 35.

17. Henry Scott Stokes, "Japanese Decide to Permit Export of Military Technology to the U.S.," The New York Times, January 15, 1983, p. 8.

18. Geoffrey Murray, "Will 'Pacifist' Japan Export Arms to the U.S.?" The Christian Science Monitor, March 15, 1983, p. 13.

19. United Nations Association of the United States of America, Issues Before the 38th General Assembly of the United Nations (New York: UNA-USA, 1983), p. 90.

20. Krishna Kumar, "Social and Cultural Impact on Transnational Enterprises: An Overview," in Krishna Kumar, ed., Transnational Enterprises: Their Impact on Third World Societies and Cultures (Boulder, Colo.:

Westview Press, 1980), pp. 20-23.

21. Robert Girling, "Mechanisms of Imperialism: Technology and the Dependent State," in Latin American Perspectives (Riverside, Calif.), vol. 3, no. 4 (1976), p. 59.

22. See Joan Pearce, "Chronology of U.N. Discussions on Technology Transfer," in W. A. P. Manser and Simon Webley, Technology Transfer to Developing Countries (London: The Royal Institute of International Affairs, 1979), Appendix II, pp. 42-48. Also, J. Davidson Frame, International Business and Global Technology (Lexington, Mass.: D. C. Heath, 1983), p. 152.

23. Emmanuel, Appropriate or Underdeveloped Technology?, pp. 1 and 103.

24. For a contrast between the views on codes of conduct for the transfer of technology between developed and developing countries, see Howard V. Perlmutter and Tagi Sagafi-nejad, International Technology Transfer: Guidelines, Codes, and a Muffled Quadrilogue (New York: Pergamon Press, 1981), table 2:2, p. 42.

25. Several of these issues were also the subject of recommendations in the so-called Brandt Report--Independent Commission on International Development Issues, North-South: A Program for Survival (Cambridge, Mass.: The MIT Press, 1980).

26. Frances Stewart, "Technology Transfer and North/South Relations: Some Current Issues," in Joseph S. Szyliowicz, ed., Technology and International Affairs (New York: Praeger Publishers, 1981), p. 214.

27. One example will illustrate the point: In 1980 and 1981 the U.N. Interim Fund for Science and Technology for Development (UNIFSTD), financed by voluntary contributions, received 870 requests for technical assistance totaling an estimated $640 million. Its resource base to meet these requests aggregated $38 million--see United Nations Association of the United States of America, Issues Before the 37th General Assembly of the United Nations, 1982-83 (New York: UNA-USA, 1982), p. 92.

28. Harry G. Johnson, "The North-South Issue," in Karl Brunner, ed., The First World and the Third World: Essays on the New International Economic Order (Rochester, N.Y.: Center for Research in Government

Policy and Business, University of Rochester, 1978),
p. 101. A number of other essays in this collection
aim at refuting the "UNCTAD approach."

29. Padma Desai, The Bokaro Steel Plant: A Study of
Soviet Economic Assistance (Amsterdam: North-Holland
Publishing Co., 1972).

30. John V. Granger, Technology and International Relations (San Francisco: W. H. Freeman, 1979), pp. 56-
57.

31. For instance, in his study of technology transfer
from Egypt, Sagafi-nejad found that among the reasons
underlying Egypt's technology exports to other countries, "request by host country government," "request
by our [Egyptian] government," and "offer of government subsidy" underlay some one-third of the transfer
cases. Sagafi-nejad, Transfer of Technology from
Egypt, p. 38.

32. George R. White, "Transfer of Commercial Technology," in Manning, ed., Technology Transfer, pp. 196-
197.

33. Victor Basiuk, "Technology and the Structure of
the International System," in Szyliowicz, Technology
and International Affairs, pp. 219-238. See also
H.-C. de Bettignies, "The Management of Technology
Transfer: Can It Be Learned?", in Richardson, Integrated Technology Transfer, pp. 110-112.

34. This is one of the major themes in David Dickson,
Alternative Technology and the Politics of Technical
Change (London: Fontana/Collins, 1974). Especially,
Dickson tries to demonstrate how capitalist interests
maintain their dominant economic-technological con-
trol over developing countries (p. 167).

35. Paulo Bastos Tigre, Technology and Competition in
the Brazilian Computer Industry (New York: St. Martin's Press, 1983), p. 6.

36. Wen-Lee Ting and Chi Schive, "Direct Investment
and Technology Transfer from Taiwan," in Krishna Kumar
and Maxwell G. McLeod, eds., Multinationals from Developing Countries (Lexington, Mass.: D. C. Heath and
Co., 1981), p. 108.

37. Vidya N. Singh, Technology Transfer and Economic
Development: Models and Practices for Developing

Countries (Jersey City, N.J.: Unz and Co., 1983),
p. 63.

CHAPTER 6

THE CASE STUDIES

Chapters 1 through 5 touched a number of bases to establish a conceptual framework and raised several pertinent issues. Accordingly, it is appropriate here to suggest a few probing questions that could be used as an overall matrix to promote critical thinking in the analysis of each case study. This means not only the ability to organize information efficiently, discriminate among the data provided, pick out a central theme in each case, and evaluate it in terms of previously acquired knowledge, but also to recognize the role of values in setting priorities and making public policy. Taken together, the questions relate to the basic elements that involve the technology transfer process.[1]

A Conceptual Framework

Question 1. The technology transfer item (Chapters 1 and 3)

What is being transferred: A tangible product? A process? General or specific knowhow? A combination of these?

If the recipient of the technology does not possess the necessary know-how to use the technology, that capability must also be transferred in addition to any blueprints, process sheets, materials specifications, instructions, and perhaps equipment and especially personnel as well, since more than documentation is necessary for a successful transfer--namely, for a sustained technology supplier-recipient relationship.

Question 2. The technology supplier (Chapter 1)

Who owns rights to the transferred technology? Is the owner willing to make the technology accessible to others? If so, under what terms? And what form of compensation is involved?

Generally, the suppliers of technology, especially multinational corporations, combine superior management techniques, product or manufacturing technologies, worldwide research and/or marketing activities, large financial resources, centralized authority structures, and good communications systems to bring technological solutions in one geographic or product market or sector to bear on a problem or opportunity in another. But how to make them share these?

Question 3. The technology recipient
(Chapter 1)

Who will receive the transferred item from the technology donor?

Technology recipients are often the overseas subsidiaries of the technology supplier (especially in the case of multinational corporations), other private firms, government-owned enterprises, private voluntary organizations, even international organizations. The public enterprises are more likely to be dedicated to development objectives than other recipients, an advantage that may have to be traded off against less than optimal efficiency (because of incompetence, nepotism, corruption, red tape, and the like) compared to private, strictly profit-oriented, firms.

Question 4. The transfer mechanism
(Chapter 1)

What channel of transfer is being used, that is, how will the transferred item be delivered from the technology supplier to the technology recipient?

The technology delivery system may be direct, such as investments in plant and equipment abroad, where the technology supplier maintains proprietary control over the technology; or indirect, such as licensing, where the proprietor grants certain rights to a licensee to manufacture and sell products but can subject the agreement to restrictive clauses, that is, relinquishes some but not all control. These two forms of transfer are often used together.

In the case of direct investments, the most important transfer mechanism, the bias in favor of transferring turnkey, ready-to-go, operations--either because these may represent maximum profit for the transferer or because the transferee is or at least is perceived to be incapable of adding value to a less complete installation or process--tends to reduce on-the-job training opportunities for local personnel and increases the cost of transfer. In the case of indirect transfers the following problems, typical of developing countries, have been identified: Export restrictions, import restrictions, sales restrictions, excessive payments.[2]

Question 5. Technology transfer timing
(Chapters 1 and 4)

When does technology transfer and/or diffusion take place?

Generally, when the personnel of the technology recipient are taught how to use, modify, adapt, or adjust the transferred item rather than just learn by rote how to adopt or operate the equipment. Some opine that technology is not transferred until it begins to diffuse within the recipient's economy.

Question 6. The absorptive capacity of the recipient (Chapter 4)

How capable is the recipient in adopting the technology effectively?

Developing countries have poor capacities to absorb new technology because of their weak scientific-technological infrastructures. Thus, their options in selecting technologies suitable for their development needs are severely limited.

Question 7. Barriers to technology transfer (Chapter 2, 3, 4, and 5)

Does a transfer actually take place?

There are numerous potential barriers to technology transfer which may stymie or prevent it from really taking place, smoothly or at all. These barriers tend to be:

a. Economic: Excessive prices charged to recipients or the inability of would-be transferees to pay even "fair" prices. More generally, deficiencies of the international market system such as its failures to factor in externalities, whether external costs or external benefits.

b. Legal: Contractual restrictions required by the technology supplier limiting the recipient's free use or control of the technology.

c. Sociotechnical: The absence of a science and technology base--say, educational capability to enable personnel to acquire the necessary skills.

d. Cultural: The lack of intranational or transnational understanding as this bears on work-related values, factors of production, interpersonal relations, social customs, group activities, and the like.

e. Additudinal: Differences in motivation for achievement, resistance to change, and other adaptational difficulties.

f. Communications: Language barriers resulting from linguistic differences or individual thought processes.[3]

Question 8. Public policy (Chapter 5)

Why is the technology transferred?

This issue has to do with certain geographic, economic, sociocultural, psychological, political, but especially value considerations. The technology suppliers may seek geographic spread either to "beat" the product life-cycle;[4] to increase profits by seeking low-cost production sites; new or cheaper sources of raw materials; competitive advantages over actual or potential rivals in the host country or third markets; relief from stringent pollution control in the home country or import restrictions in the host country; public relations factors; political power considerations; or especially to enhance one or more values given high priority.

The technology recipients may be striving to increase their own profits, market share, improve or diversify their products, or in the case of government agencies or public enterprises seek to boost tax revenues, employment, wage levels, development goals, modernization, and so on--again, subsumed in value considerations.

Through regulations, fiscal, and other public policy instruments, home and host governments strive to help their respective constituencies achieve these goals.

118

Question 9. The technology transfer/
human values symbiotic relationship
(Part I)

To what extent is the technology transferred appropriate?

The concept of appropriateness may be scrutinized in
terms of scale, technical and managerial skills, mater-
ials/energy (assured availability of supply at reasonable cost), physical environment (temperature, humidity, atmosphere salinity, water availability, etc.),
capital opportunity costs (to be commensurate with
benefits), but especially human values (acceptability
of the end-product by the intended users in light of
their institutions, traditions, beliefs, taboos, and
what they consider the good life). Or again, the technology transferred may be appropriate from a macro
viewpoint (usually, the recipient country's development) or from a micro frame of reference (usually, the
recipient firm's capability and performance).

Obviously, these questions are not exhaustive.
Other, possibly more subjective, questions could also
be raised. For instance, how does one define a fair
price for technology? What are the criteria for appropriateness?

About the Case Studies

Each case presented here has been selected on the
premise that it illustrates one or more of the fundamental concepts, definitions, and issues in the tech-
nology/society, technology transfer/human values interface introduced above. While Part I has referred
to several mini examples and macro projects, the full-
length case studies in Part II are set mostly in micro
contexts for the following reasons: (1) They are less
commonplace and thus are less readily available in the
literature;[5] (2) their components are more manageable
and therefore easier to analyze than those of macro
projects; (3) readers can relate more closely to
things they can conceptualize better.

The diversity of the cases selected provides a
broad scan of contemporary sources in different fields.
The varying origins and lengths of the cases reflect
the range of these sources and the different levels of
sophistication which were brought to bear. The aggregate displays a balance between relatively traditional
work by professionals on one hand and dedicated but

less typical neophytes on the other.

One thesis underlying these case studies is that there is no single universal technological solution to any or all developmental or international trade problems flowing from technology transfer. Indeed, the cases belie the assumption that the conclusions derived from them can be extended in linear fashion. The dynamics in every case is complex, often subjective, and different. And since dynamics implies change, the cases argue against the existence of technological determinism because, while technological monopolies and technological dependence exist, they tend to be short-run phenomena. For in the long run technological mastery is acquired by others either through transfer or through technical change impacting on both technology suppliers and recipients.

The cases should be analyzed from a number of viewpoints in terms of the consequences that technology transfer has on the physical, economic, social, cultural, psychological, or other environments which not only act on but in turn are impacted by technology in symbiotic fashion. Look for the "successful" cases, where technology transfer is relatively well adapted to the relevant universe. Look for others where the evidence indicates either a lack of sensitivity to local history or conditions, human values and culture, economic or social conditions, mindsets, political or other constraints, or for cases that show attempts to fit the environment to the transferred technology instead of vice versa. In every instance consider which variables are enhanced as a result of the transfer and which others are sacrificed in the trade-offs involved.

Most of the case studies have been written from a particular viewpoint. They therefore include the specific perspective of their authors. There is no question that if the vantage points were different-- as in the case of the three blind men touching and visualizing the proverbial elephant--so would the descriptions, recommendations, solutions, evaluations, and indeed categorizations. For many of these are admittedly arbitrary and depend on the particular emphasis placed on them.

The cases have been grouped to follow the chapter sequence as much as possible. The texts of the case studies have been left as close to their original form as necessary to conserve their initial flavor and so that the stress placed by their sources (individuals, private voluntary organizations, and international organizations of various kinds) may remain undisturbed. When any editing of the cases has been done, this has been primarily in the interest of reducing at times excessive length or excising passages that may be overly detailed or technical or of peripheral relevance to the major issues.

FOOTNOTES

1. In the establishment of this framework I have relied in part on Leo E. Konz, The International Transfer of Commercial Technology: The Role of the Multinational Corporation (New York: Arno Press, 1980), pp. 24 et seq. (originally, a Ph.D. dissertation at The University of Texas in Austin); Smith, Japanese Technology Transfer to Brazil (originally, a Ph.D. dissertation at George Washington University in Washington, D.C.), pp. 11-15; and J. Davidson Frame, International Business and Global Technology (Lexington, Mass.: D. C. Heath, 1983), p. 73.

2. See, for instance, O. S. Arthur, The Commercialization of Technology in Jamaica (Kingston, Jamaica: Caribbean Technology Policy Studies Project, 1977), quoted in Frank Long, "The Management of Technology Transfer to Public Enterprises in the Caribbean," Technology in Society, vol. 5, no. 1 (1983), pp. 69 et seq.

3. Robert E. Brasseur, "Constraints in the Transfer of Knowledge," Focus, vol. 3 (1976), pp. 13 et seq., quoted in Smith, Japanese Technology Transfer to Brazil, p. 25.

4. As a given technology nears obsolescence in the home market, its life can be extended by transferring it to new international markets, which would again increase the supplier's return on investment. This may also be achieved by diffusing it further in the home market or abroad so that the technology may be used in new or different applications.

5. There is a large body of easily available literature dealing with major technology transfer projects. A number of them are listed in Tagi Sagafi-nejad and Robert Belfield, Transnational Corporations, Technology Transfer, and Development: A Bibliographic Sourcebook (New York: Pergamon Press, 1980). A

121

concise discussion of some better-known technology transfer cases may be found, inter alia, in Ernest Braun, David Collingridge, and Kate Hinton, Assessment of Technology Decisions--Case Studies (London: Butterworths, 1979).

Case	Chapter Number and Title	Source	Case Title	Country	Page
1	1-Technology Transfer and Technology Diffusion	Alvin G. Edgell, Peace Corps/CARE	Community Cannery and Poultry Breeding	Turkey	127
2		Medicus Mundi Internationalis	Medical Aid in Reverse	Third World	131
3	2-Human Values	Sisters of the Good Shepherd	Water for the Village of Magyikwan	Burma	135
4		Rev. Andy Doral, Maryknoll Missionaries	Barrio Imas Prostitute Rehabilitation	Costa Rica	139
5		Margot L. Zimmerman, et al., The Population Council	Sex Roles in Instructional Materials: Testing the Stereotypes	Mexico	145
6	3-Types of Technologies Transferred	Maryknoll Missionaries	Lake Victoria-Mara Windmills	Tanzania	159
7		Rev. George Cotter, Maryknoll Missionaries	Salawe Pump	Tanzania	163
8		Thomas B. Fricke, A.T. International	Rainwater Collection Tanks	Thailand	167
9		Missionaries of the Sacred Heart	Single Side Band Radio Communication System	Papua New Guinea	177
10		Technoserve,Inc.	Sugar Syrup Technology	Ghana	181
11		Beverly Emerson Donohue	Textile Visual Materials	Ghana/ Sudan	187
12		Medicus Mundi Internationalis	Tetanus Vaccination	Third World	195
13		Medicus Mundi Internationalis	The Baby Killer: Formula Milk	Third World	197

Case	Chapter Number and Title	Source	Case Title	Country	Page
14	4-Problems in Technology Transfer	Roy Lock, A.T. International	Solar Energy Devices	Lesotho	201
15		Douglas Gold-schmidt, et al., Academy for Educational Development	Two-Way Radio for Rural Health Care	Worldwide	215
16		Paul Bundick, World Bank	Technology Choice in a Fisheries Project	Unspecified	229
17		Barbara Myers, A.T. International	Papa China (Taro) Project	Colombia	245
18		National Development Service/UNICEF	Communicating with Pictures	Nepal	253
19	5-Technology Transfer and Public Policy	Paul Bundick and Robert Maybury, World Bank	A Publicly-Owned Steel Mill	Unspecified	281
20		Robert Maybury, World Bank	The Choice of Technology for Textile Production	Unspecified	299
21		Paul Bundick, World Bank	Youth Brigades: Technology Choice in an Education Project	Unspecified	315

PART II--THE CASES

Chapter 1
Technology Transfer and Technology Diffusion

The cases that follow strive to illustrate the various parameters of these two related processes. One case deals essentially with the transfer of hardware, while the other focuses on the transfer of software, specifically in the form of reverse technology transfer. Examine the physical, economic, and other changes resulting, for better or worse, from the transfer discussed in each case.

Case No. 1

Community Cannery and Poultry-Breeding, Turkey

by
Alvin G. Edgell,
The Peace Corps/CARE

This case indicates that technology transfer often consists of a mix of hardware (jars, lids, pressure cookers in the case of the cannery; chicks, feed,
and medication in the case of poultry breeding) and
software (know-how, marketing, management, imagination, initiative).

A Community Cannery in Karakoy

This northwestern Anatolian village area (in
Bilecik Province, Turkey) used to produce a surplus
of vegetables and fruits, and their prices were dis-
couragingly low at harvest periods. Nor did this
seasonal surplus do anything for the unbalanced village nutrition during other times of the year. Peace
Corps Volunteer Eric Hutchinson noted these conditions
and found ready discussions on the subject to be possible, both in the village and among the provincial
officials. The village women were generally familiar
with the notion of canning since the provincial home
economist had given a course on the subject there previously. Explorations with both villagers and officials were intensive and extended over some time, for
the evolving idea of a community cannery was complicated and unprecedented in Turkey. Eric arrived in
Karakoy in late October 1964 and the canning plant
went into full operation in the late summer of 1965.

The provincial assistant director of agriculture
took a dedicated interest in seeing the project
through. The governor himself gave keen support, and,
in fact, found some funds in the provincial budget for
construction. The home economist eagerly plunged in
and, among other things, supplied the critical training in the more elaborate processes planned, with
special attention given to the village woman who
would be the continuing operations supervisor. The
adult education director for the province was also
supportive and helped in marginal ways. The village
itself put up a sizable share of the construction

costs for the simple, efficiently designed structure.

A key man in all this planning was Turkey's then forgotten top expert on small community canneries, Professor Kemal Gokce. Eric had hunted him down in his "mad inventor's" lab in the agricultural faculty of the University of Ankara. The Karakoy project promised to enable Gokce to implement his carefully worked-out ideas--and his cause. It was Gokce who took charge of the technical designing, including the building, subtly adapted to the immediate circumstances.

There were plenty of uncertain moments along the way, including those invariably attending the funding of such projects. It was clear very soon that equipment would be a major problem. Appropriate jars and lids were not then being produced in Turkey, although a manufacturer in Istanbul had been indicating his intent to produce them for some time. CARE had been supplying American jars and lids for use in just such simple courses as had been previously given in Karakoy on the assumption that a Turkish source would soon be available and that it would be nudged along by demand generated by the courses. Appropriate pressure cookers were also not made in Turkey although smaller ones were, and the capability supposedly existed. In any case, CARE would also import cookers of proper capacity.

Although the immediate goals of the project were the optimum use of the locally grown vegetables and fruits, another important purpose underlay this project, largely conceived by Dr. Gokce and heartily encouraged by CARE. It was to be a model for widespread emulation and demonstration to potential manufacturers of essential equipment so that they might be encouraged to produce it.

The cannery project came off and it had ramifying effects in both the village and outside. Starting with Eric's house to house canvassing, village interest gradually grew into excitement. Almost all the village women were eager, and the men became proudly involved in many elements of the planning, location, and provision of land for the structure, raising construction funds, etc. An apparent side effect was the rushing to completion of quite an impressive village park-garden neighboring the cannery. An element of continuing organization, unusual for villagers, was necessarily introduced for the operation of the

cannery. Among other things, the CARE jars were to be
rationed, rented, and returned to the cannery--all
duly recorded by the trained local woman in charge.
The whole nature of the project tended to add importance and prestige to the role of the village women.

But it was outside the village that the project
promised to have the greatest effect, on a potentially
national scale. For one thing, jar and lid manufacturers seemed to have been prompted into actual production. The success and lessons at Karakoy persuaded
CARE to give priority attention to supporting similar
community canneries elsewhere, in large part to con-
vince potential local equipment suppliers--including
the until recently less responsive pressure-cooker
makers--that a market existed for appropriately
adapted wares. The vision of a proliferation of community canneries with their nutritional and potential
economic advantages, supplied entirely with Turkish-
made equipment, now seemed tantalizingly near fulfillment. Officials in Bilecik Province were so impressed
that they budgeted substantial funds towards construction of three similar community canneries in 1966.

Primarily because of this enthusiastic response
in Bilecik, a Peace Corps girl with home economist
skills, Ginny Lange, was assigned to work out of that
provincial center, giving priority attention to the
three canneries established replicating Karakoy's.
She was able to put pressure on equipment manufacturers by experimenting with their pilot production
and giving them feedback on needed improvements. After
a typically disappointing 21 months in a regular village "role," Ginny found her new activity much more
satisfying and clearly useful.

What Eric seemed to contribute most importantly
was a keen, probing imagination; dedicated, dogged
follow-up on the leads it gave him; and then unflag-
ging attention to all the coordination and communication effort that had to be kept moving--primarily
among the village, CARE, and Bilecik Province's government officials--through long and sometimes anxious
periods before the grand, festive day of the cannery's
formal dedication by provincial and Ankara officials.

Poultry Breeding in Karapinar

Karapinar in south central Anatolia was the scene
of a successful experiment involving an ostensibly
limited resource, poultry. Peace Corps Volunteers

Marvin Eash and Mike Basile were assigned to the village of Islik. After a while Marvin decided to become active in a field where he had had some experience because his father had been a large commercial poultry breeder. One villager was genuinely interested in trying out modern methods with the hope of realizing some profits. But he could not or would not put up all the risk capital. So a CARE plan was drafted to cover the cost of 200 chicks, special feed, and medication required to start the project. An interested Turkish official became involved at an early stage and every step was worked out with his assistance. After a reasonable time allowed to enable the brood to start turning a profit, the farmer was to repay his seed money to CARE so that another farmer might be advanced funds for a similar poultry project on a revolving basis.

The first venture proved to be successful. Marv and Mike did their best, and then two other farmers in the village wanted to try improved poultry raising. The Turkish official became enthusiastic at the prospects in the area and wanted to expand the project on a sound basis. So a second, more encompassing, CARE project was drawn up. From among several farmers who had evidenced an interest in modern poultry breeding, 10 were carefully selected as likely candidates for their demonstration effect. A grant was to enable each of them to purchase chicks, feed, and medication. The poultry breeding project was on its way.

Questions

1. Explain the importance of the home economics course for the fruit and vegetable canning project.

2. What technology was being transferred in the case of each project?

3. How did the ability of the Peace Corps/CARE workers to "go native" assist the technology transfer process in these cases?

130

Case No. 2

Medical Aid in Reverse

by
Medicus Mundi Internationalis

This case highlights reverse technology transfer
--the so-called brain drain--involving the medical
profession. Such reverse transfer benefits developed
at the expense of developing countries, suggesting
that technology transfer can assist privileged as
well as underprivileged countries, societies, communities.

Of the 12,000 foreign medical graduates entering
the United States each year, more than 50 percent come
from developing countries. The average cost of training a medical student in the U.S.A. is $12,650 per
year, so that the replacement cost of this annual influx from the developing countries would be in the
order of $290 million a year. In the United Kingdom,
more than half of the hospital doctors come from developing countries and the situation in some other
European countries is similar. Figures on para-medi-
cal personnel are harder to come by, but even a casual
observer can see that European hospital services depend to a considerable extent on staff from developing
countries. This enormous drain of manpower and capital invested, which might nearly equal the manpower
and material aid of all Western medical assistance to
underdeveloped countries put together, has received
surprisingly little attention up to now (except certain worries about declining medical standards in the
U.S.A.). The brain drain seems to be increasing every
year and cannot be neglected any longer. What are the
causes?

1. Some (mainly Asian, South American) developing countries train more doctors than can be employed
at present because there are not enough funds to pay
their salaries, the network of hospitals and other
institutions has not been sufficiently developed,
necessary auxiliary staff is not available, etc.

Reprinted from Concepts II with the permission of
Medicus Mundi Internationalis.

2. The training is insufficiently adapted to the local needs and medical facilities. Graduates from large, well-equipped, sophisticated medical centers feel frustrated with the simple facilities and tools of the rural hospital where their services are required and look for places where they can work as they have learned in medical school. Although a number of African medical schools have adapted their curriculum to their own needs, the overall aim still is "European standard," which too often means: trained for Europe or the U.S.A.

3. In many areas, the political situation obstructs the realization of health services equal for all the people and also in other ways forces professionals to emigrate.

4. It seems that university training alienates the students from their own culture. It creates new cultural needs, both spiritual and material, that can hardly be fulfilled in their own country and certainly not in the rural areas where the majority are needed.

The major solutions are outside the realm of medicine. But a crucial point certainly is that health services in some developing countries are unwittingly set up in too Western a style. It should be possible to create systems that are culturally and economically integrated while still providing a maximum of services which present medical science can make available. The function of the doctor in the system should be reconsidered. The training of medical students according to Western models sometimes impresses as a neo-colonialist anachronism.

Questions

1. Explain how Third World medical students transfer technology to advanced countries.

2. To what extent is the brain drain a cultural, a professional, and a political problem?

3. Can you think of other resources, besides technology, being transferred from developing to developed environments?

Chapter 2
Human Values

These cases illustrate the fundamental premise of the entire work--that technology shapes society as it is shaped by it. Notice how human values, ultimately reflecting people's perception of reality and involved on both sides of the technology transfer process, have an especially determining role in the success of each transfer.

Case No. 3

Water for the Village of Magyikwan, Burma

by
Sisters of the Good Shepherd

This case focuses on the successful transfer of basic, infrastructural technology, hardware and software, in a very simple village setting where water has both physical and symbolic meaning. The fact has led the technology users to ponder over the underlying values of the technology suppliers in an uncommon way.

Background

The remote village of Magyikwan ("Magic Wind" in Burmese) is 73 miles north of Rangoon, Burma's capital, but can only be reached by walking a mile along a rough footpath or by riding an ox-cart. Approximately 1,000 Burmese villagers live there representing about 300 families. Five Burmese Good Shepherd Sisters occupy a two-story barn-like convent not far from the little church. In addition to their evangelical endeavors among adults and children, the Sisters are the only group striving to bring some modicum of social development to this highly depressed area.

The poverty of the village is very great. Its inhabitants have no cooperative farms or any food production. At harvest time the men hire themselves out at other villages as seasonal day laborers, but during the summer months they have no work whatsoever. Sister Rose, superior of the Good Shepherd nuns, explained:

> The people have no food in their homes, really nothing. When mealtime comes, everyone (adults and children) goes out looking for wild rice or some other scraps to eat. There is a kind of wild, bitter bean which they also collect. The people never have enough to eat, but they share what little they have.

In addition, the villagers are all in perpetual debt to the moneylenders, and it is impossible for them to provide their families with the basic needs of food, shelter, or clothing.

The main reason for this state of extreme destitution is that there simply is no water in the village. The people have no water for growing crops and of course no sanitary facilities, water for bathing, or washing clothes. The water they use for drinking is foul. It comes from a large hole dug in the ground and partly filled with dirty rainwater. They use this supply as drinking water until the end of January when the hole dries up. Sister Rose explained that from January to April each year there is no water in the village. She noted:

> ...we have a well for surface water, but large amounts of both sulfur and iron in the water make it unusable. Thus, the people must go with an ox-cart and buy water from another village three miles away. At times this village will not sell water if its own supply is low. Even when the neighboring village does sell water, each family can only buy one jarful and bring it back here with the ox-cart.

It is easy to see that nothing can be done about the food problem until the serious water problem is solved. There is good land in the village, and food could be grown under the guidance of the Sisters, but only after a village water supply is secured.

The Project

In 1976 Sister Rose tried to solve the water problem by the installation of a very simple manual pump. She obtained 4,000 kyats (approximately U.S. $670) from "The Little Way," a small British charitable organization. With these funds she was able to have a manual pump and a cement tank installed in the village. The pump brought clean, clear water through a plastic pipe from a pool of water 120 feet deep. The water would be stored in the tank and the villagers would get their supply by dipping their receptacles into the tank.

Of course, this delighted the people. They would wait in line for hours, sometimes all day, with pots in hand in order to secure some drinking water for their families. However, because of poor quality and constant use, both the pump and the plastic pipe leading to the pool of water deteriorated after three years of service in late 1979. By the end of January

136

1980 the villagers were left with nothing: The waterhole dried up and the pump was inoperative.

Because of this predicament, Operation HELP, Inc.,
provided $200 to patch things up on a temporary basis
by inserting a smaller plastic pipe inside the larger
one and by replacing the rings in the pump.

What is needed, however, is good-quality equipment, a power pump and an air compressor, that will
endure wear and tear so that larger quantities of
water may become available not only for drinking but
also for sanitary purposes and for crop irrigation.

The Sisters pointed out that the presence of the
pump not only helps the villagers in a material way
but that it also epitomizes the concept of selfless
service. Indeed, even the Buddhist members of the
community (to whom water is a kind of symbol of life)
have become interested in learning more about the
Sisters and why they live in the village and make the
people's problems their own.

Questions

1. What is the relationship between the technology
transfer agents (Sisters of the Good Shepherd) and the
technology transferred in this case? Is this a common
or an unusual relationship?

2. Which human values do (1) the transfer agents and
(2) the transfer users exhibit?

3. When a community exists at near-destitution levels
as in Magyikwan, can transferred technology, no matter
how simple, represent more of a quantum leap toward
development than at higher standards of living?

Case No. 4

Barrio Imas Prostitute Rehabilitation, Costa Rica

by
Father Andy Doral,
Maryknoll Missionaries

> This case deals with one social side effect of transnational technology transfer--prostitution--consequent on the establishment of large-scale foreign-run banana plantations in a Central American country. Also, the transfer of other technology in the form of rehabilitative skills to correct the situation, suggesting the value-laden nature--both good and bad--of technology.

The Background

The immediate purpose of the Barrio Imas project is to provide for a work and study center designed for use by ex-prostitutes. These women are now living in a poor and decaying neighborhood in the port city of Golfito, Costa Rica, located in the heart of a string of large banana plantations owned by the United States-based United Brands (formerly, the United Fruit Company). This center is not an end in itself but is intended as a setting for both productive work and social interaction. Such activity will in turn allow for the restoration of personal dignity and social identity, thereby mitigating the marginal status of these women in the broader community. Specifically, the project will strive to achieve the following goals:

1. To raise the personal and family status of the ex-prostitutes of Barrio Imas;

2. To alter the structure of the local economy so that prostitution may no longer be regarded as a necessary source of income;

3. To reduce the social stigma experienced by these women; and

4. To break down intra-community animosities.

Before the program, activities, and organization

139

of the center are described, two fundamental questions must be raised. First, what are the socio-economic origins underlying the problem of prostitution along Costa Rica's Pacific coastline? Second, how does the building of a work-and-study center address this problem so that a solution for the poverty and marginalization of prostitutes may at least be sought?

An attempt will be made to answer those questions within the context of a brief discussion of the relationship between prostitution on one hand and the growth of "La Compania" (United Brands) on the other.

Any consideration of social problems along Costa Rica's Pacific Coast must begin with an analysis of the development of these large agricultural enclaves--the banana plantations. In the closing decades of the 19th century, the building of railroads throughout Latin America coupled with the greatly increased world demand for bananas precipitated the penetration of United States capital into the Caribbean area. The early years of these plantations were chaotic as competition for land among a large number of small companies intensified.

In 1899, the United Fruit Company (United Brands) was formed as a conglomerate of 22 smaller enterprises. Increased efficiency in production made possible by economies of scale and the coordination of previously diffused railroad systems heralded a new period of expansion in Honduras, Panama, Colombia, Guatemala, and Costa Rica with United Fruit purchasing huge tracts of land along the Atlantic Coast. During the 1930s, blight swept through these plantations and United Fruit switched banana production to the Pacific Coast, using its old "fincas" (farms) for the cultivation of African palm, with Golfito the leading banana port.

The formation of these agricultural enclaves led to a dual economy in Costa Rica. The industrial sector, oriented to the production of manufactured goods for the domestic market, revolves around the capital, San Jose. The agricultural sector, on the other hand, focuses on the "Compania" and primarily on the production of export commodities. The relationship between these two sectors is not antagonistic but mutually supportive. Too, United Brands has retained significant influence on national policy despite recent trends involving the nationalization of some of the plantations.

140

Let us now mention the nexus between dual economy on one hand and the solution of social problems on the other. In Costa Rica, such social problems as alcoholism, prostitution, marginalization, or even poverty cannot be treated as a whole because the problem varies depending on whether one speaks of the urban metropolis or the rural enclave. Despite such differences, however, national policy has tended to be defined only by the requirements of the urban context, leaving the problem as it exists in the enclave untouched. The poor people in the enclave regions are systematically neglected by social reforms which serve only to alleviate pressures of poverty surrounding the more affluent metropolis. Underlying this neglect is the fact that political interest is not focused on the needs of the masses. In the specific case of prostitution the consequences of this state of affairs are now described.

Enclave Prostitution as a Social Problem

On the one hand, the enclave economy is firmly supportive of traditional female roles (wife and mother) attached to the home and the family. However, in creating a labor market that is dynamic and seasonal in its operation, United Fruit undermined the role of the family as a social unit, indeed, constantly placing it into question.

Government programs have generally focused on attempts to rehabilitate prostitutes through housing or family planning programs. While this may be sufficient for urban, "industrial," prostitution, such programs do no more than blur the clarity of the problem as it exists in enclave port towns. In Golfito, local culture distinguishes between card-carrying and occasional prostitutes. While rehabilitation may have the effect of shifting a prostitute from the first category to the second, rarely will she be moved out of the profession altogether. In failing to challenge the basis of production in the enclave and in allowing the continued dominance of large corporations like United Brands, such programs may change the appearance of the problem but not its reality. Reforms may succeed in winning votes for politicians but not in altering the necessity of prostitution as a source of income for families without a male head of household in the enclave. This is precisely what happened in Golfito, the largest banana exporting port.

In 1964 the Hotel Romero, which had been used as

141

a house of prostitution since 1946, was condemned by
the Costa Rican Board of Health. But alternative housing was not provided at the time, and a large number
of prostitutes continued living and "working" in the
building. In January 1970 a rent strike was called,
protesting what had become definitely substandard living conditions. In March of that year, in the heat of
election promises made during the preceding months,
the government finally agreed to send a team of inves-
tigators down to Golfito. Their report confirmed the
findings of the Board of Health made six years earlier.
A government committee was then formed to deal with
the problem, defined by the committee as a housing
issue rather than as one of social marginalization.
In May 1970 a decision was made to allocate funds and
resources for the construction of 25 pre-fabricated
housing units with labor donated by the families of
the community. In June 1972 the units were completed
and the families began to move out of the Hotel Romero.
Barely three months later the building collapsed,
killing three of the remaining residents.

Rehabilitation

In 1973 the community began construction of the
work and study center. The following year, 15 Barrio
Imas women obtained certificates for their completion
of sewing courses, scheduled to be repeated. But following a change in government and new policy priorities, construction of the center came to a halt in
1975. The Barrio Imas Committee has repeatedly attempted to find other sources of financing in the
country but with little success. In a recent interview, the "Madrina" (president of the Barrio Imas Committee) explained the difficulty clearly: "The government no longer cares about our votes. The rest of
the town forgot about us long ago. And the Church
cannot do anything. We have to try outside."

The Barrio Imas Project attempts to cover two
basic programs:

1. To decrease the reliance on prostitution as a
source of income for fatherless families by initiating
a local clothing industry;

2. To substitute communal for family support in
the context of a wide variety of social activities.

This effort would fill the void in the government's program by addressing the problem of

142

prostitution at its roots.

The Barrio Imas Community Project has arranged community activities (such as problem-solving sessions), educational activities (such as adult reading and writing courses), economic activities (a sewing or toymaking workshop to provide work and income for the women and their families), recreational activities (such as sports, films, drama, dances, community festivals), and motivational activities (counseling). The Community Council which runs the project has issued progress reports beginning on November 1, 1978, attesting to the success of the project.

Questions

1. Identify an enclave community similar to a Costa Rican banana plantation owned or operated by a foreign multinational corporation--say, an Arabian-American Oil Company community in Saudi Arabia. What would be the effects of oil technology transfer on the American employees and their families living in such a community? On the company's native employees and workers? On the wider recipient community? On Saudi government officials?

2. Do you consider the prostitution rehabilitation program appropriate from the viewpoint of the local culture in Barrio Imas?

3. Think of some other examples where the harmful social effects of technology transfer are offset or remedied and some where they are not.

Case No. 5

Sex Roles in Instructional Materials:
Testing the Stereotypes

by
Margot L. Zimmerman, et al.,
PIACT, Seattle and Washington, D.C.

This case describes how in Mexico, a traditional
society, literature on the use of a drug for the
treatment of a children's disease represented adult
males in atypical sex roles. And while the message
was found to be effective, sensitivity is urged in the
use of representations to promote transnational--read,
transcultural--technology transfer.

Reprinted with the permission of the Population Coun-
cil from Margot L. Zimmerman, Maria Elena Casanova,
Dianna P. Stern, and Sarah S. Auman, "Sex Roles in
Instructional Material: Testing the Stereotypes,"
Studies in Family Planning 13, No. 8/9 (August/
September 1982), pp. 262-270.

145

146

Special Report

Sex Roles in Instructional Materials:
Testing the Stereotypes

Margot L. Zimmerman, Maria Elena Casanova,
Dianna P. Stern, and Sarah S. Auman

For the past few years, PIACT, the Program for the Introduction and Adaptation of Contraceptive Technology, has been helping to develop culturally appropriate printed materials that rely mainly on illustrations to explain correct contraceptive use to illiterate and semiliterate family planning acceptors and potential acceptors. We have undertaken these projects because we believe that nonreaders, who constitute 75–80 percent of the adult population in some developing countries, need access to accurate and reliable information to which they can refer when making decisions and following instructions. Our experience has shown that such information is most effectively and economically acquired through the use of the print medium, in conjunction with personalized oral instruction in individual or small group settings.

On the basis of its experience in preparing printed support materials containing pictorial instructions, PIACT has been asked to prepare materials to help nonreaders properly use various health products. PIACT's first health-related pamphlet, on how to mix and give oral rehydration salts (ORS) solution to a child with diarrhea, was designed in Mexico. In 16 pages, this pamphlet imparts such messages as: if your child has diarrhea, take him/her to a clinic where ORS is available; wash your hands before emptying the packet of ORS into one liter of clean water; give a little of the solution to the child following each bowel movement with loose stools; do not give the child other medicine at the same time he/she is taking the ORS treatment; continue to give the

child food; if you are nursing the sick child, continue to do so even though he/she is also being given ORS.

The pamphlet was well received in Mexico and is currently being used by the Ministry of Health in a national program to reduce the incidence of infant dehydration. But in the United States there was some criticism over the fact that no men appeared in the illustrations contained in the pamphlet. It was suggested that PIACT was reinforcing stereotypical sex roles since the illustrations show only women involved with the sick child. Was PIACT perpetuating a bias that such activities can only be performed by females?

PIACT staff, both in the United States and in Mexico, was skeptical about adding a male to the pamphlet. "Everyone" knows the aura of machismo surrounding the Mexican male, who considers his role to be family provider while his spouse handles all domestic chores and is the nurturer of children.[1] We were told by social workers and communication personnel that both women and men would ridicule any material that suggested that men could also be involved in child care activities. And if such were the case, encouraging male involvement would only discredit the material and, by implication, the product whose use we were promoting—in this case, oral rehydration salts.

The Study

We decided to undertake a small experiment on sex stereotyping in informational materials for illiterates and semiliterates. Despite the initial skepticism, PIACT began with the hypothesis that portraying nontraditional sex roles for men would not interfere with comprehension of the technical information contained in the pamphlet on how to prepare and use oral rehydration salts (ORS). Using two pamphlets on ORS, identical in every respect except that one

Margot L. Zimmerman, B.A., is Program Officer in charge of IE&C projects, PIACT, Canal Place, 130 Nickerson Street, Seattle, WA 98109. Maria Elena Casanova, is Manager, Materials Development Projects, PIACT de Mexico, A.C. Dianna P. Stern, M.P.H., is Administrative Assistant and Sarah S. Auman, M.S.W., is Research Assistant, PIACT.

showed a man (father figure) in all the pictorial messages relating to care of the child with diarrhea, we proposed to test both versions of the pamphlet for individual message comprehension, information recall, acceptability, and pamphlet preference in a small comparative study. While our major emphasis would be on interviewing women who were representative of the recipients of these pamphlets in any ongoing ORS program, we also decided to compare the reactions and preferences of men with young children and health providers when they were shown both versions of the pamphlet.

The study was designed to determine whether (1) the potential users of the pamphlet would object to or reject the idea of the father as purveyor of health care to his sick child; (2) the presence of the father figure was, in any way, an obstacle to message comprehension; and (3) people would prefer a pamphlet with both the mother and father caring for the child with diarrhea, or one that showed only the mother caring for the sick child.

Methodology

In order to have two identical versions of the pamphlet, an artist copied the original printed version and then prepared a revised version that included the male "caretaker." The male "caretaker" was added to almost every page of the revised pamphlet. In most frames, he is shown either assisting or accompanying the woman. In some frames, he is shown working alone—for example, pouring the ORS solution, feeding the child, and dressing the child following his stay on the "potty." The presence of the father—either with the mother or substituting for her—is the only difference found between the two pamphlets.

We divided the women into two groups and gave each a separate test so we could compare two different learning modalities: (1) individual message comprehension as a woman examined the pictures and (2) recall of facts after a fieldworker had reviewed the pamphlet with her, message by message. When testing the men, we decided not to ask them to recall specific tasks, since we felt the entire experience would be new to them and we were reluctant to have them feel threatened or put on the defensive. We were more interested in their comparisons between the two pamphlets—did they notice the differences and, if so, which did they prefer and why.

Two hundred subjects from low socioeconomic sections of rural and suburban central Mexico were included in the study.[2] We wanted our sample population to correspond to the educational level of the sample population used in developing the original ORS pamphlet. Thus, women and men with more than six years of schooling were eliminated from our sample during preliminary interviews (see Table 1).

Other criteria for the selection of study subjects were that subjects be the direct or indirect users of ORS (i.e., mothers and fathers of young children or health workers) and that the mothers either be married to or living with a man.

Interviews with mothers and fathers were carried out in private homes; those with health workers were conducted at their place of employment—Social Security clinics in suburban areas near Mexico City. Each respondent was approached individually. After a brief introduction, potential subjects were told enough about ORS to provide some perspective on what would ensue. Persons who met the criteria for inclusion in our study were interviewed using standard questionnaires prepared to secure data that would enable us to compare the comprehension of the different messages among the different groups.

Two PIACT de Mexico staff members, experienced social workers, were trained by the project director in the interviewing and recording techniques required for this study. Following a day of role-playing, several pilot interviews were conducted to pretest the interview and data collecting instruments, to clarify any doubts, and to correct any possible misunderstandings among the interviewers.

Four different interview forms were designed to correspond to the four groups in this study—Group A, B, C, and D—described below:

Group A

Sixty rural or suburban mothers of young children reviewed the ORS pamphlets without any text. Our design called for 30 subjects to go through the original pamphlet, showing only the mother, while the other 30 would view the revised pamphlet, showing both parents. But as we were analyzing the data, we realized that someone had miscounted and we actually had 28 viewing the original pamphlet and 32 viewing the revised pamphlet. As a PIACT interviewer went through the booklet, page by page, and asked each woman to explain what she saw in the picture about ORS and to interpret the messages on how to use ORS, the PIACT observer recorded the woman's verbal and/or nonverbal comments, interpretations, explanations, and expressions pertaining to each pictorial message.

This was the only group that saw pictures without the text, as we wanted to look solely at their comprehension of pictorial messages. Once the respondents had viewed and interpreted all 16 messages in the pamphlet, they were shown whichever version they had not seen originally, were asked to note any differences between the two versions, and to explain

148

TABLE 1 Educational levels and family sizes of mothers and fathers viewing original and revised pamphlet

	Group A: women (N=60)		Group B: women (N=80)		Group D: men (N=30)	
	Number	Percent	Number	Percent	Number	Percent
Schooling						
No schooling						
Original	6	21.0	11	27.0	2	13.3
Revised	8	25.0	7	18.0	4	26.7
1–4 years						
Original	19	68.0	28	68.0	11	73.3
Revised	21	66.0	31	80.0	9	60.0
5–6 years						
Original	3	10.7	2	4.9	2	13.3
Revised	3	9.4	1	2.6	2	13.3
Number of children						
2 or less						
Original	7	25.0	10	24.5	3	20.0
Revised	8	25.0	13	33.3	4	26.7
3–5						
Original	16	57.1	23	56.0	9	60.0
Revised	19	59.4	22	56.4	8	53.3
6 or more						
Original	5	17.9	8	19.5	3	20.0
Revised	5	15.6	4	10.3	3	20.0

which version they preferred and why. They were also asked whether they had ever received help from their husbands at home in taking care of a sick child, and if so, to describe the kind of child care help they had received.

Group B

In another test for comprehension and recall following an oral message reinforced by a detailed drawing, 80 rural and suburban mothers of young children reviewed *only one* version of the ORS pamphlet. Again, our subgroups did not turn out to be exactly even. Forty-one women were shown the original pamphlet, while 39 viewed the revised version. Women in both subgroups were from a low socioeconomic background, with most having a maximum of three years of schooling. This group (as well as Groups C and D) viewed the same pictorial pamphlets used by Group A, except that these women also saw a line or two of simple Spanish text. The PIACT fieldworker went through the pamphlet with each woman, page by page, describing and explaining the message portrayed in each illustration. The women were encouraged to ask questions and to make comments; many did so.

Following a detailed explanation (as contained in the illustrations) of when, how, and why to use ORS, the pamphlet was put aside and the woman was asked seven questions based on the information contained in the pamphlet. Since the information

provided in both versions of the instructional pamphlets was identical, all 80 women were asked the same questions, in the same order. Responses were noted by the PIACT recorder, who also noted each woman's comments, facial expressions, and any other reactions.

Group C

Thirty health workers (all of them female nurses) from urban and periurban areas around Mexico City were shown both versions of the ORS pamphlet; 15 of them viewed the original version first, while the remaining 15 viewed the revised one first. The health workers received some general information about the pamphlets and how they were used. They were then asked to comment on the material and to give their perceptions and preferences based on their knowledge of the target audience with whom they work. Their responses and comments were noted by the PIACT recorder.

Group D

Thirty fathers of young children were told about ORS and why the pamphlets were being developed; they were then shown both pamphlets. Again, half were shown the original pamphlet first, while half were shown the revised version first. The men interviewed came from low socioeconomic backgrounds and earned no more than 1,500 pesos (under US$70) per

149

week. Most had completed between two and three years of schooling.

They were given as much time as they wanted to "study" each pamphlet, but once they had finished looking at one they were not permitted to refer back to the first one while examining the second. When they had completed their viewing, the fathers were asked for their opinion on the pamphlets, their preferences, comments, and so forth. Their responses were noted by the PIACT recorder.

Results

The data gathered during this study and summarized in this section yielded two important findings. First, the subjects' comprehension and rate of recall of the technical health care information in the pamphlets were not affected in any significant way by the presence of the father figure being portrayed in nontraditional sex roles. This finding upheld our original hypothesis: portraying new or nontraditional sex roles for men does not interfere with comprehension of the technical information contained in the ORS instructional pamphlet. Second, the majority of both male and female subjects who were exposed to both versions of the pamphlet preferred the version that included the father figure and portrayed him as assisting the mother in caring for the sick child.

Group A: A Test for Individual Message Comprehension

Of the subgroup (N=28) that viewed the original ORS pamphlet and were asked to explain the meaning of the pictorial messages, only a few had trouble verbalizing what the illustrations meant. The same was true of the women (N=32) who "read" from the revised ORS pamphlet. We discovered that several women in both subgroups had trouble with two of the messages: (1) "Don't give the child any other medicine" (while he/she is taking the ORS solution), and (2) "Continue giving the ORS solution to the child whenever he has a loose stool, even during the night." However, their inability to understand was due to their misinterpretation of the meaning of the symbols used, and was unrelated to the presence of either parent in the illustration. To symbolize night, we had used a burning candle. Many women, in both groups, thought the candle was lit because the child had died. Many others said the candle was lit in honor of a saint to ensure the recovery of the child. We had used an "X" when the nurse was explaining that no other medication should be given to the child while he was taking the ORS solution, but found that

TABLE 2 Number of women correctly explaining selected pictorial messages without text or previous instruction, in original and revised pamphlets

	Original		Revised	
Message[a]	No.	%	No.	%
The child is crying; he has diarrhea.	28	100	32	100
The diarrhea continues, and the parents are worried.	27	96	32	100
The parents take the child to the hospital.	28	100	32	100
Only one liter of the ORS solution should be given in 24 hours, divided into 5 cups each.	27	96	30	94
One cup of the medicine should be given to the child every time he has a loose stool.	28	100	32	100
Continue giving him the medicine even during the night.	21	75	21	66
If the child is being breastfed, this must not be interrupted during ORS treatment.	28	100	32	100

[a]Translated from the original Spanish.

several women did not know what the "X" symbol meant.

As can be seen in Table 2, we found no significant differences in individual message comprehension (without explanation or instruction) between the women who based their interpretation on the original pamphlet and those who did likewise with the revised pamphlet. The messages chosen for inclusion in Table 2 are those in which the father figure plays a prominent role in the revised version.

After explaining one of the pamphlets, each woman was shown the second pamphlet and asked to look it over. When finished, the PIACT interviewer asked each woman questions about her pamphlet preference and about her personal experiences regarding assistance from her husband in child care.

Forty-one (68 percent) of the 60 women interviewed preferred the revised version of the ORS pamphlet; of these women, 27 (45 percent) gave reasons relevant to the subject matter of the study, such as: "The presence of both parents makes the picture more complete" (N=5); "So that the men will see that not only the woman can take care of the children, the two should share in caring for the sick child" (N=12). The other 14 women gave reasons unrelated to the subject of the study (e.g., "It explains the drawing better," etc.).

Eighteen (30 percent) of the 60 women preferred the original version of the ORS pamphlet, but only eight of those gave reasons related to the study: "The mother is the one who cares for the children" (N=4); "The father is not at home to take care of the children" (N=4). The other ten gave reasons unrelated to the study, including such false perceptions as "the paper is thicker" or "the drawings are clearer."

One person out of the entire group of 60 thought that both pamphlets gave the same message, that the two pamphlets were very good, and that it did not matter which one was used.

Thirty-eight (63 percent) of the women said their husbands do help them when the children are ill. Fifty-four (90 percent) of the women agreed that the husband should help his wife when a child is sick, because: "Both parents have the same obligation toward the children" (N=27); "It shows his interest in cooperating with her—this she likes" (N=17); "Two heads are better than one" (N=4); "They should help whenever they can" (N=2); "I would like him to help me, but he doesn't want to" (N=2); "It took both of them to have the child, so they should both help" (N=3). Six (10 percent) of the women did not agree that the men should help with their sick child, because: "The men have to work" (N=4); "It isn't the same when the father cares for the sick child as when the mother does" (N=1); "They never worry about the children" (N=1).

Group B: A Test for Recall

This group was given a test for comprehension and recall of instructions imparted, to determine whether the presence of a male figure interfered with the women's ability to remember the specific ORS instructions. Of the 80 women interviewed, 41 were shown only the original version of the ORS pamphlet, while the other 39 viewed only the revised version. Responses to some of the questions asked after the pamphlet had been explained are as follows. (For a list of all questions and number of respondents giving correct answers, see Table 3.)

What would you do if your child had diarrhea? Of the women who examined the original version, 28 (68 percent) answered correctly that they would first take the child to the "hospital";[3] four women (10 percent) said they would give him some medicine in the form of a syrup for the stomach; five (12 percent) said they

TABLE 3   Number of women correctly answering questions testing for recall of instructions in original and revised pamphlets

Questions	Original (N=41)		Revised (N=39)	
	No.	%	No.	%
What would you do if your child had diarrhea? (Answer given in pamphlet: Take child to clinic.)	28	68	29	74
How often should ORS be given in one day? (Answer: After each instance of diarrhea.)	36	88	36	92
How much ORS solution should the child be given? (Answer: One liter, in 5 cups.)	35	85	36	92
Should you stop breastfeeding your infant while he/she is taking ORS? (Answer: Mother should continue breastfeeding.)	36	88	38	97
Should you feed a child who is taking ORS? (Answer: It is important to feed the child while he/she is taking ORS.)	38	93	37	94
Is it alright to give the child other medicines while he/she is taking ORS? (Answer: Do not give the child any other medicine.)	34	83	37	94
How do you prepare ORS solution? (Of five steps given in the pamphlet, number remembered):				
5	1	2.5	2	5
4	12	29	12	31
3	14	34	14	36
1 or 2	14	34	11	28

151

would give him tea with a tablet of "terramicina"; and four (10 percent) said they would give the child a tablet and plenty of liquids. Of the women who examined the revised version, 29 (74 percent) answered correctly that they would take the child to the "hospital"; five women (13 percent) said they would give the child some tea; three (8 percent) said they would buy some medicine at the pharmacy; and two (5 percent) said they would "stop feeding the child milk."

How often should you give the ORS solution to the child in one day? Of the women who examined the original version, 36 (88 percent) answered correctly "after the child had a loose stool"; three women (7 percent) said every five hours; and two (5 percent) said a small amount every two hours. Of the women who examined the revised version, 36 (92 percent) answered correctly that they would give the child a cup of the solution every time he/she had a loose stool; one woman said she would give the child the liquid once every 24 hours; and two women (5 percent) said they would administer the solution four times daily.

How much ORS solution should you give the child in one day? Of the women who examined the original version, 35 (85 percent) answered correctly "one liter in small portions"; two women said five liters daily; and one answered "a little every day." Three women did not answer the question. Of the women who examined the revised version, 36 (92 percent) answered correctly "one liter in small portions"; one woman said "one cup"; another said "one-fourth of a liter"; and a final woman said "one jug full."

Is it alright to give the child other medicines while he/she is being treated with the ORS solution? Of the women who examined the original version, 34 (83 percent) answered correctly that medicines should not be mixed; seven (17 percent) said they could still receive other medicines while on ORS. Of the women who examined the revised version, 37 (94 percent) answered correctly that no other medicines should be given; two (5 percent) said it would be alright to mix medicines.

How would you prepare the ORS solution? Of the women who examined the original version, 14 women (34 percent) mentioned one or two of the five steps correctly; 14 (34 percent) correctly mentioned three steps; 12 (29 percent) mentioned four steps; and only one (2.5 percent) was able to recall all five steps correctly. Of the women who examined the revised version, three (8 percent) correctly mentioned one step only; eight (20 percent) mentioned two steps; 14 (36 percent) remembered three steps; 12 (31 percent) knew four steps; and two women (5 percent) remembered all five steps correctly.

As Table 3 indicates, the presence of a male figure did not interfere with the ability of the women who had seen the revised version to recall the instructions and messages. The two subgroups were quite similar in their ability to provide correct answers to the questions. As a matter of note, the women who were shown the revised version had slightly but not significantly higher scores on recall than the women who were tested on the original pamphlet.

All the women interviewed were very interested in ORS, since they do not like to put their children in the hospital. Some of them mentioned various home remedies they give their children with diarrhea, such as herbal teas or soda water with orange juice, but mentioned that physicians usually prescribe some other medications along with the solution for diarrhea.

None of the women who saw the revised pamphlet commented on the fact that both father and mother were taking care of the child. Only one woman, as indicated by a surprised facial expression, reacted in any way to the father's preparing the ORS solution.

Group C: A Test among Health Workers

Of the 30 health workers (female nurses) interviewed, 15 analyzed the original version of the pamphlet first and then the revised version. The other subgroup of 15 did the reverse. The results from this group were as follows.

Twenty-six of the health workers (87 percent) said the two pamphlets explained the preparation and administration of the ORS solution equally well. Ten of these 26 realized that the father was portrayed in one of the pamphlets helping the mother to care for the sick child. Nevertheless, they affirmed that the two pamphlets were the same in all other respects. The four remaining health workers thought the original pamphlet was clearer, and emphasized that it is in fact the mother who is with the child most of the time.

When asked which of the pamphlets they would use to show families how to prepare and administer ORS correctly, 20 of the 30 health workers (67 percent) chose the revised pamphlet. Seven of them said, "It is more complete because it shows both the father and the mother"; five said, "The father should have the same obligation as the mother, and this pamphlet encourages the man to assist his wife"; and two said, "It is important that the father participates, since the children's problems concern both parents." The remaining six health workers mentioned reasons

152

for their preference that were unrelated to the study and based on false perceptions, such as "larger drawings," "better photographs," and "more pages."

Eight of the health workers (27 percent) preferred the original version of the ORS pamphlet, with six of them saying that "It is the mother who spends more time with the children." The other two health workers gave irrelevant reasons based on false perceptions, such as "thicker paper" or "brighter drawings."

Group D: Determining the Perceptions of Men

Thirty men were interviewed to determine their reactions to both versions of the ORS pamphlet, as well as their preferences. All men were shown both pamphlets and asked to view them slowly, after which they were asked several questions. These questions and the men's responses follow:

Is there any difference between these two pamphlets? If so, what are they? Eighteen of the men (60 percent) said they did notice a difference; 15 of these men pointed out that one pamphlet showed both parents. The other three men noted disparities unrelated to the study ("type of paper," "the wall is peeling," etc.).

Which pamphlet do you think is better and why? Twenty-one of the men (70 percent) preferred the revised version; 16 of these gave one or more of the following reasons: "It is the obligation of both parents to care for a sick child"; "That is how the man ought to help his wife"; "It is more complete because both parents are present." Five of the men who preferred the revised version of the pamphlet gave reasons unrelated to the study.

What do you think of a father helping to take care of his sick child? Twenty-five men (83 percent) gave positive answers to this question, saying one or more of the following: "The man should help his wife because it is their mutual responsibility to care for a sick child"; "It is important that the father helps"; "They should both help with the child." Five men (17 percent) did not feel the father should help in caring for his sick child, saying either, "The man is never home," or "The father does not have time."

Have you ever helped in caring for your sick child? If so, in what way? Twenty-one men (70 percent) said they had helped their wives care for their children when they were ill. The ways in which they had helped included: "I've taken sick children to the doctor"; "I have given the child his medicine at night"; "I do whatever is necessary to help"; and "I clean the child when he has diarrhea."

Discussion

Little or no research has been done on illiterate or semiliterate men's attitudes toward helping with sick children in developing countries. As a population, this group has been grossly neglected by the conventional approaches to health in developing countries. There has been an increasing awareness of the "changing roles of women" but almost a complete lack of data on the roles of men. According to Safilios-Rothschild, "While the roles of women are changing unevenly and at different speeds in different societies, the roles of men are also undergoing changes which may not necessarily correspond to changes taking place in the women's roles."[4] Safilios-Rothschild has studied the role of the family in development and concluded that, among the poor, family organization is characterized by flexibility. "The criteria for distributing resources and responsibilities within families in developing countries do not coincide with Western criteria of development needs and requirements. Nor do sex and age, especially among the poor, determine the roles people play in the family."[5] Following her line of reasoning, it is perhaps not so surprising that the predominantly low-income Mexicans who comprised most of our study population responded favorably to seeing a man in what middle- and upper-class men and women in Mexico and in developed countries consider to be traditional female roles. And similarly, our group of nurses, also educated to favor traditional role models for men and women, were less enthusiastic about the presence of a man in a book on child care than were the mothers and fathers from low socioeconomic levels.

According to Stokes, recent studies have found that men have much more interest in family planning and a willingness to practice it than they are given credit for.[6] Although an earlier study found that Mexican men wanted large families and viewed their wives as mothers, housewives, and as a means of sexual satisfaction,[7] Stokes says that these conclusions give a one-dimensional image of the Mexican male. Most men surveyed by Stokes rejected machismo behavior in *other* men, especially the attitude that having a large family makes someone feel more manly. The vast majority of those he interviewed also believed that "in general, one should limit the number of children," citing the economic benefits of smaller families as well as concern for the future welfare of their existing children. Stokes also found that poverty, not male chauvinism, often shapes men's attitudes.

In Nigeria, where care of seriously ill children is often relegated to fathers, men are very receptive to learning how to administer oral rehydration fluid in the home. By so doing, they have found they can avoid the distress of sharing in the child care tasks when their children become seriously ill. This usually involves buying expensive drugs and carrying the child great distances to seek help.[8]

In a paper on fertility decision-making within the household, Hollerbach suggests that, within Latin American lower- and working-class populations, experiments with programs designed to educate and involve men, utilizing male health personnel and community workers, seem particularly worthwhile.[9] Studies in these communities indicate that the male peer group, rather than the family, may be the source of identification and social approval, as well as the major source of influence on fertility norms and behavior. Many of the sociological and economic evaluations of development projects have overlooked the structure, dynamics, and behavior patterns of the family. Hollerbach feels that more basic research on the poor in developing countries is needed so that data explaining observed patterns of change and the effects of development efforts to alleviate poverty can guide planners and programmers.

A case study to illustrate this need was done in an area of Kenya where the kwashiorkor rate among children dropped from 55 percent to 39 percent just 18 months after the fathers of the affected children were made aware of the nutritional problem and became involved in its solution, using locally available resources.[10] Prior to the active participation of the fathers, who decide what foods are to be given to children in this area of Kenya, years of conventional nutrition education at maternal and child health clinics had been directed toward the mothers, who actually feed the children. Had the clinic programmers first gone to the villages to identify the decision-makers—hence the potential problem solvers in this case—years of wasted effort could have been avoided. What ultimately moved educational efforts out of the clinic and into the villages was a simple statement made by one of the fathers during a village meeting with district health officials to discuss the continuing high rates of kwashiorkor: "If this [kwashiorkor] is such a problem, why hasn't it been brought to the attention of the elders? When wild pigs destroy our crops or other calamities threaten village life, the elders are always consulted. . . . Certainly we should be consulted on matters that involve the lives of our children. . . . We don't get pregnant, so why would we go to the Women's Clinic [MCH clinic] to learn of this problem?"

This Kenyan study of the extent to which men want and need to be involved, coupled with PIACT's recent findings, again point out the mistake develop-

ment planners make in focusing almost exclusively on women and, perhaps unconsciously, using stereotypes to define roles women should play when the subject is health—especially child health.

Conclusion

Contrary to general belief, the subjects in this PIACT study did not reject the idea of the father helping the mother to care for the sick child; nor did the presence of the father figure interfere with comprehension of the ORS instructions. These unexpected and collective results have significant implications for future materials-development work in Mexico. They also have implications for instructional and motivational communication endeavors throughout the developing world.

Our hypothesis, that portraying nontraditional sex roles for men in ORS instructional pamphlets does not interfere with comprehension of the technical information contained in the pamphlets, was upheld in this study. More importantly, we found that the majority of male and female subjects who were exposed to both versions of the ORS pamphlet preferred the version in which a father figure appeared performing nontraditional sex roles. Collectively, these findings demonstrate that in our study population the simultaneous presentation of nontraditional sex roles with technical instructive messages is both compatible and advantageous. The simultaneous presentation is advantageous because it has the potential for reaching a wider audience. Widescale distribution and use of this version of the pamphlet throughout the population might mean communication to a much larger audience, as men might be attracted to and/or included in the target audience. Furthermore, this larger audience of both parents would also receive messages on two different levels: on the surface, technical instruction on correct health care will be transmitted; and then just beneath the surface, is the behavioral message that it is acceptable—and perhaps even desirable—for fathers to play a key role in the care and nurturing of their children. A male presence in these pictorial messages might, in fact, contribute toward modifying behavior and defining new sex roles for men. If this were the case, it would mean doubling the human resources available to children for informed and appropriate home health care in times of illness.

Whether or not the goal of a health education communications project is to transmit both instructional and behavioral modification messages, our subjects preferred the version of the ORS pamphlet in which both messages were presented. And, pref-

154

erence in itself is a powerful element in motivation for learning and changing—whether it be at the cognitive level only or at both the cognitive and behavioral levels. Therefore, when program planners are motivating men to take a larger role in caring for the family's health, they should consider exposing men to materials similar to the ORS pamphlet that contained a father figure.

What PIACT has learned from this study is likely to alter our approach to the development of instructional print materials that are primarily pictorial. On all health, nutrition, and family planning topics that are not of necessity sex-specific, men will be included in the initial phases of materials development to ensure their full participation and subsequent representation in whatever final products are developed. Whenever appropriate, males will be portrayed helping women and children in messages regarding parental assistance to children. In all family planning materials, PIACT has been careful to portray husbands and wives discussing the various methods of contraception, and we usually show the husband accompanying his wife to the family planning clinic. But in our health materials, especially those involving children, we have tended to reinforce society's bias and have shown only women feeding and caring for children.

As a result of this study, we feel that the use of ORS pamphlets in which men are portrayed should reach a larger audience, transmit technical information, and might even assist in breaking the perpetuation of traditional stereotypic sex roles for women. When it is time to reprint those materials, we recommend that serious consideration be given to substituting our revised version for the original version currently used in Mexico. We encourage all program personnel engaged in producing family planning, health, and nutrition materials to include a role for fathers as a natural adjunct of their "parenting." We also recommend that other organizations in other countries conduct studies similar to this study to determine whether our findings are transferable to other societies whose child health care materials

make women solely responsible for the care of their young.

References and Notes

This study was made possible by a grant from the Ford Foundation.

1 P. Huston, *Message from the Village* (New York: Epoch B. Foundation, 1978), pp. 109–142; The Population Council and PIACT de Mexico, *Family Planning in Mexico*, December 1979; E. Folch-Lyon, L. de la Macorra, and S.B. Schearer, "Focus group and survey research on family planning in Mexico," *Studies in Family Planning* 12, no. 12 (December 1981): 409–432.

2 Interviews were conducted in the suburban areas of Barrio Norte, Cuidad Netzahualcoyotl, Santa Fe, and Naucalpan; and in the rural villages of Ayotla, Texcoco, Boye, and Villa Guerrero.

3 Our study population used the word "hospital" to refer to any clinic or outpatient medical facility.

4 C. Safilios-Rothschild, "The demographic consequences of the changing roles of men and women in the 80's," in *Proceedings of the 1978 Conference on Economic and Demographic Change: Issues for the 1980's* (Liège, Belgium: International Union for the Scientific Study of Population, 1978), pp. 4.1.2-1.

5 C. Safilios-Rothschild, "The role of the family in development," *Finance and Development* (December 1980): 1–4.

6 B. Stokes, "Men and family planning," *Worldwatch Paper* No. 41 (December 1980): 7–10.

7 The Population Council and PIACT de Mexico, cited in note 1.

8 Anonymous, "Fathers' Club," *League for International Food Education Newsletter* (November 1981): 2.

9 P. Hollerbach, "Power in families, communication and fertility decision-making," Center for Policy Studies Working Paper No. 53 (New York: The Population Council, January 1980), p. 29.

10 J. Balcomb, "Enlisting fathers in the fight against malnutrition," *UNICEF News*, Issue 92, Number 2, 1977, pp. 9–11; Anonymous, "Fathers help fight malnutrition in Kenya," *League for International Food Education Newsletter* (November 1981): 1.

Questions

1. Was it surprising that criticism of the original version of the pamphlet (showing only the mother involved in child care) should have originated in the United States?

2. How is it that portraying nontraditional sex roles for men in the ORS pamphlet did not reduce the credibility of the technical information contained in it?

3. What are the implications of the unexpected survey results comparing the two versions of the ORS pamphlet in the preparation of instructional materials for transfer to other societies?

Chapter 3
Types of Technologies Transferred

These cases suggest that the inclusion of the concepts of intermediateness and appropriateness into the process of technology transfer mandates examining the relevant economic factor inputs in the case of intermediate technology and even broader sociocultural and psychological criteria in the case of appropriate technology. Such frames of reference lead to the obverse issue of misapplied, or inappropriate, technology. But because all these terms are situational, relational, and dynamic, and since trade-offs are always involved, the dividing line between appropriate and inappropriate technology may be difficult to establish, anticipating the problems concerned with the making of technology transfer assessments.

Lake Victoria-Mara Windmills, Tanzania

by
Maryknoll Missionaries

This case highlights the interface between a
given technology transfer--windmills--on one hand, community needs and attitudes on the other. Notice the
conditions and timing that tend to make such a transfer appropriate in this case.

Background

The Mara region of Tanzania lies on the southeast
shore of Lake Victoria, the largest in Africa. The
people of the region are farmers. They herd small
flocks of cattle, goats, and sheep, and cultivate private and cooperative plots of corn, millet, rice, cassava, and vegetables. They live in homes of mud with
grass roofs. One out of three families may own a bi-
cycle or radio.

The Mara region is crossed by two broad rivers,
the Mara and the Mori. Both rivers flow year round
into Lake Victoria. Despite this potential water supply, the Mara people suffer from severe droughts.
Since 1973 the rains have been inadequate. Thousands
of their cattle have died. Their crops have withered
unharvested in the fields. Their children have felt
the effects of malnutrition.

For 30 years the Maryknoll Fathers have been involved in the Mara region in a wide range of agricultural and human development projects. They have a
good knowledge of tribal languages and a basic understanding of the people. In the past 15 years, the
Maryknollers in Tanzania have initiated 10 small windmill projects. They brought sample windmills from
Australia, Kenya, and the United States, and built
others using Land Rover transmissions and available
parts. Some of these projects were successful and
continue to serve the people today.

Maryknollers are now planning a large-scale project which will make water-pumping windmills available
to the people of the Mara region. The project will

involve maximum community participation. It will be based on the previous windmill projects and on information from the feasibility study (see below).

By now the government has completed its program of moving the farmers into cooperative villages. This policy has brought about changes in the peoples' thinking and has made them aware of new possibilities. This is therefore an ideal time to introduce a system of windmills which will bring much needed water to the farms and villages of the Mara people.

Below is a brief overview of the proposed windmill project, and a detailed description of the feasibility study which must precede it.

Project Overview

The Maryknoll Fathers of the Diocese of Musoma perceive the introduction of windmills as providing much needed service to the Mara people. Four factors have encouraged their thinking:

1. The Mara region has the necessary natural resources: An abundance of lake and river water, steady winds, and large tracts of low-lying black soil along the shores of Lake Victoria.

2. The inadequate rainfall in recent years has sharply reduced the food supply of the Mara people.

3. A critical shortage of machinery parts and the high cost of diesel fuel make power pumps impractical in rural Tanzania.

4. The Tanzanian government's policy of self-reliance stresses appropriate technology.

Although windmills are simple mechanisms, experience has shown that they are fragile and costly to maintain in rural Africa. Among traditional people technical skills are limited, and machinery parts are difficult to replace. Therefore, the introduction of windmills demands thorough preliminary study and careful planning.

If the findings of the feasibility study are favorable for the Lake Victoria-Mara Windmill Project, a team of two windmill specialists and three African trainees will undertake the project for a period of three years. They will work with the Maryknoll

Fathers, taking advantage of their knowledge of local customs, their rapport with government officials, and their acceptance by the villagers.

It is expected that the entire project will be turned over to the Tanzanian government at the end of the three years. In that case the government will assume responsibility for:

- Four windmills to supply water to villages;
- Four windmills to supply water to irrigated vegetable gardens;
- A team of trained Tanzanian windmill technicians;
- An established line of supply to provide new windmills and windmill parts;
- Windmill tools and workshop facilities in the villages;
- Future requests for windmills from cooperative villages.

It is expected that the eight windmills involved in this project will be models for other Mara windmills and that the support services established by the project will build and maintain many future windmills.

Questions

1. What are the factors--specifically, of a sociocultural nature--that are especially likely to make the Mara windmill project a success?

2. Consider this statement from context: "They (windmill specialists and trainees) will work with the Maryknoll Fathers, taking advantage of their knowledge of local customs, their rapport with government officials, and their acceptance by the villagers." Is such a context likely to make the transfer of windmill technology appropriate?

3. The mention that "some of these projects (windmills) were successful...." suggests that others were not. Make up some scenarios where the latter would apply.

Case No. 7

Salawe Pump, Tanzania

by
Father George Cotter,
Maryknoll Missionaries

This case highlights the appropriateness of a specific type of water pump transferred to a Tanzanian village: It is simple, inexpensive, and its installation involved the community and won its acceptance. While its advent made changes in community life, it did not cause disruptions in the existing value system and could thus be considered economically and socioculturally appropriate.

"Hyenas eat human excrement and then drink water from our waterhole."

"I saw a dog cooling itself in our waterhole. It was so diseased that all its hair had fallen off."

The Wasukuma people of Salawe in Tanzania made these complaints at a village meeting in 1968 when Father George Cotter invited them to build a cement well and install an iron pump.

Though their superstitions forbade them to tamper with the waterhole and their money was limited, they agreed to try building a well. Each household contributed a share of the $20 needed for materials.

At dawn the following Wednesday the men of the village arrived at the waterhole with hoes, buckets, shovels, and a wheelbarrow. Father Cotter came by Land Rover with bricks, pipes, tools, and a cement cover.

The waterhold was a spring-fed pond whose water was always dirty. Cattle waded in it while women washed their cooking pots. Girls laundered their clothes on its edge, and everyone bathed in it. Still, they drew their drinking and cooking water from it.

The villagers worked all morning bucketing out the water, scraping away the mud, and deepening the eye of the spring. They worked steadily, and when a

163

catfish squiggled in the mud they splashed and fought for it.

At noon a column of women crossed the fields balancing pots of steaming cornmeal. While the villagers ate their lunch and relaxed, Mbuke, a woman with child, told of her visit to the pre-natal clinic. The nurse had informed her she was suffering from amoebic dysentery and that it could affect her child. Although the nurse gave pills for treatment, she advised the woman to be careful of dirty water.

Mbuke tried to retell the nurse's warning: "The animals swimming in the dirty water are so tiny that a person's eye cannot see them, but if you swallow them they reproduce and multiply and fill up your insides. And that's sickness."

An old man answered, "Aw, Mbuke, I'm a Christian. I don't believe all that superstitious stuff." Everybody laughed and the men went back to the pond to lay the bricks for the wall.

Towards evening the men had put each row of bricks in place, backed them with clean sand, and filled in the rest of the pond with dirt. After they laid the cover on the well and installed the pump they laughed with delight and surprise and began to pump out water. "The pump works!"

After that 50 neighboring villages installed Salawe pumps in their waterholes. Each village paid the cost of the materials and did the digging and building work. Father Cotter provided transportation, tools, and "know-how."

The villagers are growing accustomed to clean water. One family said, "When we travel to visit our relatives we carry along our drinking water." Several families remarked: "Our children are not sick as frequently as they used to be. We will never go back to drinking dirty water."

Questions

1. Speculate on the differences which the installation of the Salawe pump probably made to the lifestyle of the villagers involved.

2. Explain how the villagers could have been made to accept this technology even though "their superstitions forbade them to tamper with the waterhole."

3. List all the human values and cultural dimensions that came into play in transferring the Salawe pump technology to the village.

Case No. 8

Rainwater Collection Tanks, Thailand

by
Thomas B. Fricke,
Appropriate Technology International,
Washington, D.C.

This case highlights the transfer of a water collection and storage technology system to a developing
country with a long dry season. Also, the efforts
made to adapt the technology to the new setting so as
to make it appropriate.

Problem Identification

The origin of the Community-Based Appropriate
Technology and Development Services (CBATDS) Rainwater
Collection and Storage Project in northeastern Thailand can be traced to the formative discussions, consultations, and brain-storming sessions held in Bangkok during 1979. The key decision-maker, Meechai
Viravaidya, secretary general and chief conceptualizer
for the Population and Community Development Association (PDA), and Dr. Malee Sundhagul, then director-
designee of CBATDS, used consultants for feasibility
analyses and program priority deliberations but primarily drew upon their extensive technical and development planning expertise. Khun Meechai is a prominent, unconventional, and controversial development
economist, and Dr. Malee is a highly respected micro-
biologist, a dynamic program coordinator with solid
scientific credentials. After determining that the
northeast should be the priority area for development
activities and incorporating the inputs of other expert advisers, Dr. Malee and Khun Meechai selected
water supply and sanitation as a priority emphasis.
Besides the obvious need from a public health point of
view, water supply and sanitation seemed to be a promising entry point for field research, development,
and dissemination for the newly formed CBATDS.

Abridged and reprinted with the permission of
A.T. International.

The Survey

In January 1980 Dr. Sam Johnson of the Ford Foundation became involved in program support discussions in Bangkok. Ford's interest in PDA and CBATDS had been spurred by a combination of familiarity with PDA's promise, professional relationships, and the formation of CBATDS with flexible support from Appropriate Technology International (ATI). Doctors Malee and Johnson held extensive consultations on their mutual interest in establishing a program to develop a field technology adaptation and dissemination methodology, and water supply in particular attracted priority attention. A specific program for a pilot field experimentation scheme was designed and Ford awarded CBATDS a $200,000 grant to implement it.

```
                          Table 1
              Profile of Northeastern Thailand


Number of provinces                                    16
Total population                               15,792,825
Area (km²)                                        185,156
Population density (persons/km²)                     85.3
Average annual rainfall (mm)                        1,250
Average annual days of rainfall                        95
Average farm size (hectares)                            4
Average farm household income (baht)               10,280
Average family size (persons                            7
```

PDA/CBATDS field staff and technical volunteers stepped in to give the program shape, direction, and momentum. CBATDS district supervisors and family planning volunteers are a known quantity in several northeast Thailand provinces, and their legitimacy and presence had been established for several years in rural communities. Formal and informal consultations with local government officials, farmers, and government extension workers from the ministries of health and agriculture ensued in early 1980 and were followed by the design of a rudimentary and practical water resources survey format. CBFPS (Community-Based Family Planning Services) volunteers and supervisors were utilized by CBATDS's still small staff for information gathering and needs and demand assessments regarding water supply. The survey was intended to be a representative sampling which would assist in prioritizing follow-on research and development.

The survey identified technology issues and developed a short list of potential technical responses to the severe drought conditions facing people, crops, and livestock in the northeast. From the standpoint of availability, quality, capital costs, maintenance and operation, and convenience, the options were as follows: Surface water catchments, piped water, wells (with or without pumps and casements), rainwater collection and storage, and small-scale water treatment. One of the key CBATDS technical volunteers, Brian Bruns of the Peace Corps, carried out a detailed socio-economic and cost-benefit analysis of these options (Bruns 1981). Bruns reports that the following responses were most often elicited from villagers:

- Drinking water had a higher priority than water for other uses, owing to the saline conditions of ground-water;

- Health considerations, i.e., means of reducing pathogens via boiling, filtration, treatment, etc., were not mentioned as a priority concern;

- Rainwater collection where feasible and adequate in quantity was an attractive alternative to the ubiquitous open pond catchments for many villagers;

- Sophisticated, high capital cost, or community-scale systems held little attraction and were not likely to succeed unless sizable subsidies were made available; and

- Villagers required technical, logistical, and financial support to reduce the risks and generate adequate incentives for a water supply program to make any inroads into rural communities.

While the surveys were being performed, CBATDS began the pilot prototype activities. Normally such activities would not be undertaken until the survey was complete; however, in the case of CBATDS, the surveyor-facilitators had an established, credible presence in the communities they entered, and flexible logistical and material support with clear guidance from headquarters was forthcoming. Another factor often mentioned by the CBATDS staff and local participants was that the presence of outside catalysts, at first primarily foreign volunteers, provided added perspective and momentum to this project.

The choice of the particular technologies--bamboo-reinforced concrete tanks and wire-reinforced water jars--by a sizable and growing number of villagers appeared to be in response to the pilot introduction scheme elsewhere in Thailand. Villagers were impressed by the convenience, construction methods, and financing scheme used in that project. During village discussion and survey sessions, rainwater tanks or jars spontaneously appeared as a viable option for many participants. Demand, acceptance, and availability apparently merged throughout the crucial period from mid-1980 through mid-1981 when the effort picked up momentum as information and experience related to the implementation of these village-based catchment systems spread within and between villages in the target area.

The rainfall regime and ground and surface water conditions in the villages where tanks have been successfully implemented are remarkably similar, thus contributing to a successful fit and effectiveness between the technology and its users' requirements. Annual averages for the region of 1,365 millimeters are rarely achieved in these areas, and rainfall is concentrated in infrequent torrential rain, thunderstorms, and cyclonic disturbances during the rainy season which generally lasts from May to September. Minimal amounts fall during the seven-month dry season, thus placing severe stress on human, animal, and plant populations. Villagers generally obtain drinking water from open catchments on vacant or community land if it is not overly muddy, turbulent, or loaded with organic matter. Also, rainwater is collected and stored or water is purchased and brought back home from towns or cities where commercial sources are available. The water table in most of the areas visited ranges from 6 to 15 feet, but the water is generally judged too saline for human consumption.

Dissemination Strategy

CBATD's dissemination strategy and the concepts upon which it is based are explained in CBATDS's documents (CBATDS's Annual Report, 1981). When the specific technology has been determined by field staff and ratified by Bangkok administrators, CBATDS builds upon the social mobilization strategy of the CBFPS, which is based on the assumption that communities are capable of perceiving and solving their own development problems, given adequate guidance and motivation. It includes the transfer of authority for local program

170

planning, operation, and day-to-day evaluation at the community level. Such community oriented programs tend to avoid the cultural insensitivities and communication failures plaguing the delivery systems.

CBATDS views marketing as a critical component of its efforts to disseminate appropriate technologies throughout Thailand. The surplus generated from marketing activities increases the economic base of ongoing programs and of farmers. Our aim is to encourage local people to recognize their innate ability to provide for their own betterment. We provide technical and financial assistance, but ultimately it is the people themselves who create development. Investing their labor, their savings, and their local resources will bring self-sustaining prosperity over time.

CBATDS's dissemination strategy for its rainwater tanks project stresses an optimum combination of cooperative work, soft loans (revolving credit fund), and logistical and technical assistance. A view of how this process took shape in northeastern Thailand follows.

The Pilot Project

The village of Ban Lan, Ban Phai District, Khon Kaen Province, where the first tanks were built, presents an instructive account of the technology adaptation and dissemination process. In the course of discussing water supply problems during the water resources survey, ex-CUSO (Canadian) volunteer engineer Paul Grover, who is fluent in Thai, built up a friendship with a group of local farmers and craftsmen. The group included members of the village council, other influential members of the village, and two men with experience in construction. One of these was a young mechanic who became a focal point of a discussion group which was formed after the meeting with CBATDS's staff. Ban Lan is a primarily agricultural village of 300 families located on a paved road linking Ban Phai with Mahasarakham. The electric grid was just extended to this village in 1978, and it could be considered low to moderate income relative to the rest of the surrounding area. Animal traction for agriculture, charcoal as a cooking fuel, and traditional raised wooden houses with tin roofs predominate in Ban Lan.

The discussion group held several meetings which led to a determination by 20 members to commit

171

themselves loosely into a work group. The design used was based on one developed by the Sanitation Department of the Ministry of Health, which lends out the steel forms needed to make the tanks to groups and individuals interested in self-help. Villagers had examples of both functioning and non-functioning rainwater tanks available at schools in surrounding villages.

Twelve tanks were constructed in a second effort four months later, 15 in a further installment in late 1981, and preparations for a more massive construction program of 30 tanks at one time were underway when this report was being prepared. Several other enterprises were formed as a consequence of CBATDS-sponsored activity in Ban Lan village. Two experimental biogas plants were constructed in late 1980, but no momentum towards further dissemination has become evident. Additionally, CBATDS procured low-interest private loans to enable the young mechanic, at whose house the first tank was built, to establish a metal fabrication shop. This workshop is now self-supporting and produces metal cement-mixing forms (among other things) for CBATDS field operations. Since the original 51 tanks were built in seven villages in a span of six months in late 1980 and early 1981, the project has been institutionalized as a special unit within CBATDS.

CBATDS staff and the local construction crew went through this first experience in October 1980 directly in a trial and error fashion. They decided to pour all six tanks in stages of two ring forms per tank until each tank was completed. Ultimately, all six tanks performed their tasks of collecting and storing rainwater with only minor difficulties (some faucets had to be replaced, one tank had to be replastered owing to poor quality concrete). The finished tanks, ranging in size from 6 to 11.3 cubic meters, served as a strong stimulant to generate demand, instill confidence in prospective users, and increase community pride.

Successively, as interest rates rose, CBATDS supported further tank construction with its loan funds, an increasingly capable work force evolved (dubbed "technicians" by Bangkok staff), and a more streamlined and selective promotional system developed. Construction practices and tank size became standardized, designation of participants became less ad hoc, and participation was assured through incentives (lower costs to users and fines assessed on families who failed to contribute labor). CBATDS staff began going through formal local governmental channels to obtain endorsements for construction efforts. Work crews were taught scrupulously to measure the amounts of sand, stones, and other material brought to worksites to insure honesty. This proved to be difficult, since open confrontations are anathema to many Thais. (Table 2 summarizes the method CBATDS uses in disseminating this technology.)

This developed infrastructure now includes three small trucks for transporting formwork, cement, and other supplies from one building site to another. Revolving loan fund collection, storage, disbursement, and accounting are also based in the field offices.

Tungnam I

The project initiated in Ban Phai District has been repeated in over 70 villages in eight subdistricts of Khon Kaen and Mahasarakham Provinces. A grant from Agroaction of West Germany for $350,000 (₿7 million) for the construction of 1,000 tanks was initiated in May 1981. The project was called Tungnam I (tungnam means "water tank"). In early March 1982, two months ahead of schedule, the 1,000 planned-for tanks had been constructed, and Agroaction was preparing to finance a follow-up funding program, Tungnam II, totalling approximately U.S. $1,080,000 (₿20 million) earmarked for the construction of 2,500 to 4,300 water tanks in the course of two years, from June 1982 to June 1984.

Basic Description

CBATDS adapted and improved designs and construction practices for rainwater cisterns previously existing in northeastern Thailand. These collection and storage vessels are primarily of two types: Large volume (6 to 11.3 cubic meter capacity) bamboo-reinforced concrete cylindrical tanks; and wire-reinforced thin-membraned concrete semi-spherical jars (1.4 to 2 cubic meter capacity). These are intermediate technologies, combining traditional and conventional (modern) construction methods. During the remainder of this report, reference will be made primarily to the large volume tanks, since they are at this time the more prevalent and popular of the two options. They are not currently commercially available whereas the jars are. (Over 1,000 tanks and approximately 50 jars had

173

Table 2

Chart Summarizing Software Aspects of CBATDS's Technology Transfer

I. Project Preparation	II. Entering the Village	III. Implementation	IV. Follow-up
Inform provincial government officials about project and seek permission for implementation.	CBATDS staff meets with village health workers, family planning volunteers, and headman to explain project.	Metal forms and construction tools are moved into the village.	Postcards are left with tank owners (any difficulties with the tanks can be reported easily to CBATDS.
CBATDS initial survey of village domestic water resources/community leaders identified for village committee.	CBATDS staff, health worker, headman, and family planning volunteer work together locating interested villagers.	Supply shops are contacted and materials ordered with the assistance of the village committee.	Necessary repairs are made.
Contact district public health officials—explain project and seek cooperation.	A meeting is held among interested villagers. Policy is determined.	Formation of construction teams from individual households.	Continued motivation through the village committee.
District health officials and CBATDS work together locating suitable villages.	Village technicians are located and invited to a training session about rainwater tank construction.	Preparation of ground and bamboo.	Continued construction through sponsorship program.
Appropriate indigenous materials are located (i.e., bamboo, banana stalks).	A village committee is established to help oversee construction activities, insure village cooperation, and assist with distribution of contracts.	Construction of tanks.	Monitoring and evaluation.
CBATDS hires technicians, clerical staff and acquires office space/purchases tools and construction forms.			

This chart is taken from Introducing Hardware Technology with a Soft Touch by Ed Anderson and Bob McKeon (mimeographed), August 25, 1981.

been installed by April 1, 1982.)

The two exhibit similar characteristics. They both use cast-in-place construction with portable shuttering (formwork) and are easily installed by inexperienced labor (assisted by local skilled technicians) in a relatively short period of time. The containers are compatible spatially, culturally, and functionally with most existing rural Thai homes, which generally have corrugated galvanized iron roofs with some form of rainwater collection via gutters and drains for one or more roofslopes. Portable reusable forms make for ease of construction.

Questions

1. How might the order of priorities have been different if the five-point list relating to water availability and use were drawn up by a Western source instead of the Thai villagers?

2. Explain the meaning of the word "culturally" in this quote from context: "The containers are compatible spatially, culturally, and functionally with most existing rural Thai homes."

3. With reference to technology transfer assessment, comment on the "follow-up" column 4 in Table 2 entitled "Chart Summarizing Software Aspects of CBATDS's Technology Transfer."

Case No. 9

Single Side Band Radio
Communication System, Papua New Guinea

by
Missionaries of the Sacred Heart

This case highlights how even a strictly Western
technology--a radio communication network--can be entirely appropriate in an island developing country
setting under the conditions described, thereby helping to achieve some of its developmental goals.

Soon after their arrival in 1947, the American
Missionaries of the Sacred Heart recognized the urgent
need for a communication network among the 22 island
stations that make up the Catholic Diocese of Kavieng
in Papua New Guinea in the South Pacific. Indeed, the
diocese, which corresponds to the New Ireland and
Manus districts of Papua New Guinea, comprises 85,000
square miles of which 4,500 square miles represent the
land area while the rest is ocean. Accordingly, an AM
(amplitude modulation) network was installed and radio
voice contact has been maintained among the stations
twice a day ever since. This communication system has
played an integral part in the lives of the inhabitants. For example, it is used to put doctors in direct contact with out-station clinics, to request air
transport for patients needing immediate hospitaliza-
tion, to keep tabs on the movement of small craft
bringing supplies or moving children to and from
school in what are often rough waters.

The land area includes three major islands (New
Ireland, New Hanover, and Manus) and seven smaller
island groups (Duke of York, Feni, Tango, Lihir,
Tabar, Bipi, and Western). The distance of the
smaller island groups from the major islands and between each other varies from 50 to 70 miles, except
for the Western group which is several hundred miles
distant.

The population of these islands totals about
80,000, with some 25,000 living on the smaller island
groups. Roads are few. Air and sea travel are restricted and often hazardous.

There are four clinic-hospitals with resident doctors. The rest of the islands and villages have minimal if any nursing care provided by the missionary sisters and/or lay practical nurses.

Primary education is now available in the main villages of the islands. But there are only three central high schools to which the youngsters are taken by boat twice a year.

The present radio communication system is manned by the Missionaries of the Sacred Heart and the local people they have trained. Used for the benefit of individual communities without regard to religious affiliation, the network consisted of transceiver (transmit-receive) units operating on amplitude modulation (AM). However, because a single side band (SSB) system makes many more channels available for use and thus makes more radio broadcasts possible than a double side band, and in a context of growing traffic and the government's wishes to make its system compatible with the networks of other countries, as well as by virtue of international agreements relating to the control of the air waves, the AM system was gradually phased out and the SSB system, incompatible with AM, was introduced for the use of private operators by December 1, 1978, and for air traffic by 1982.

This changeover has been prompted by a request from the provincial government in Papua New Guinea to Reverend Alfred M. Stemper, D.C., bishop of Kavieng, which read in part:

> The Government Departments in this Province [New Ireland] rely heavily on the Mission radio network to contact more inaccessible areas. The Mission radio network provides a more adequate communication than can be provided by the Government....It is definitely Government policy that the present obsolete sets be phased out and the Mission network give the Government an opportunity of contacting many areas quickly in times of emergency, which is of benefit to the overall administration of the Province. Any assistance which can be given to the Catholic Mission would benefit the communications of the New Ireland Province generally.

The SSB units were manufactured by Codan Pty.,

Ltd., of Australia. The sets, which cost about $30,000,
proved useful to those who made the changeover--gov-
ernment and mission posts on the mainland of Papua New
Guinea. Suitable for tropical conditions, these units
are almost maintenance-free but can be serviced locally
by the missionaries, who were also qualified to install them. Indeed, the diocese of Kavieng is the
legal owner of the radio network.

The Missionaries of the Sacred Heart have played
an active role in what is now Papua New Guinea since
1884. The country obtained its political independence
in September 1975. Its future is promising but for
now it needs technical and other assistance.

Questions

1. Why is radio communication, a Western technology
import, so appropriate to Papua New Guinea?

2. Why were the radio sets purchased from Codan Pty.,
Ltd., Australia, especially appropriate in this particular case?

3. What other communication medium, also a Western
import, is used and is equally appropriate to this
particular setting?

Case No. 10

Sugar Syrup Technology, Ghana

by
Technoserve, Inc.,
Norwalk, Connecticut

This case highlights how a transferred technology is diffused by duplication or replication. Also, how an alternative sugar-producing organization was established to make it appropriate to a new setting.

For inspiring words of wisdom, for encouragement in times of trouble and sometimes just for fun, Ghanaians read daily, like a horoscope, the sign inscribed on their mammy trucks. The sign which has encouraged Technoserve after a decade's hard work in the field of self-help enterprise development in Ghana reads "Good never lost."

As part of its work in Ghana, Technoserve established two sugar syrup factories, Promase Limited in the Eastern Region and Alanfam Limited in the Central Region. The Technoserve development team in Ghana--now made up entirely of Ghanaian development experts--was both surprised and delighted to see two more factories being built and successfully managed in rural Ghana. This was achieved without local or overseas aid and without any direct management assistance from Technoserve. Rather, it was done by the most natural development process known to man--by copying. This fascinating development could not have come at a better time for Ghana.

In 1981 both Ghanaian government sugar plants at Asutsuare and Komenda were near bankruptcy because of mismanagement, inefficiency, and a lack of essential spare parts and machinery. For sugarcane farmers this meant a threat to their major outlet. For distillers, canning, and pharmaceutical plants and certain key chemical plants, it signified a lack of raw materials for their operations. A Ghanaian government spokesman noted that no fewer than 30 large industries and 40

Abridged and reprinted from the Technoserve News-letter with the permission of Technoserve, Inc.

medium-sized industries were dependent on sugar, alcohol, and molasses. The replication of Technoserve's sugar syrup technology did not solve all the problems, but it indicated how, in a small and effective way, modest rural-based industries, when they are seen to be well-established and viable, will be quickly duplicated in other parts of the country and will thus strengthen its economy.

Copy No. 1: Agomeda Sugar Syrup Factory

In 1979 a retired army colonel visited the Promase Limited sugar syrup factory at Nnudu, Eastern Region, Ghana. At that time it was under Technoserve management and the intensive training of key factory staff was in process. The colonel discussed the process of converting sugarcane into syrup using the bagasse (cane fiber) for fuel for the furnace and was impressed by its simplicity and ingenuity. He returned to his own village of Agomeda some 20 miles from the Promase factory in an equally good sugarcane growing area and began to draw up plans for his own Agomeda mill.

His next step was to visit the other factory under Technoserve management, Alanfam Limited, and from there he hired the local production manager on a temporary basis. The production manager began to direct the construction work using, as at Promase, 55-gallon oil drums as water-cooling tanks for the diesel engine and even the same metal fabrication workshops to make the boiling pans. The cane crushers were bought in Accra. The open factory structure itself consisted of wooden pillars and tin roofing sheets. By the end of 1979, the Agomeda sugar syrup factory was a reality. Since then, 20 to 25 workers of both sexes have been employed at the Agomeda plant and as a result whole families have benefited from the income generated there. The factory has also created a new ready market for cane produced by local farmers. This plant is close to the government sugar factory, Asutsuare, mentioned earlier.

Copy No. 2: Kissi Sugar Syrup Factory

At first sight, this factory would appear to be practically a duplicate of the Alanfam or Promase factories, but it is more than that. It is a copy of the very concept of copying the Alanfam factory! The Kissi sugar syrup factory was built and is now run by the same production manager who was hired in 1979 on a temporary basis by the retired army colonel in Agomeda. While working on the Agomeda factory, he realized the possibility of replicating such a plant in a sugarcane area. Borrowing some capital from a local business woman, he began construction of a sugar syrup factory once again--this time for himself. He chose a site 50 miles west of Alanfam Limited (Mankessim) in the town of Kissi, which is only three miles from the second large government sugar mill, Komenda. Because of the problems faced by the Komenda mill, many sugarcane farms had remained unharvested for two to three years. In 1981 the factory was completed. As in the case of the mill at Agomeda, 20 to 25 jobs were created and a new, urgently needed market for the sugarcane of local farmers was established.

The early years of developing and refining the Technoserve sugar syrup technology at the Promase and Alanfam factories in Ghana were not easy. There were times when it seemed useless to persevere. But by 1981, with the additional two mills actually established, a juice-crushing plant operating in Mankessim, and who knows how many more in the planning stage, the mammy wagon's encouragement to persevere had come true --"Good never lost."

Promase, Ltd.: A Model for Small-Scale
Intermediate Technology Involving
Sugar Cane Products

Introduction

A small-scale sugar products industry has developed in Ghana. The impetus for this development has been the desire by the Ghanaian authorities to indigenize the production of sugar and related items to alleviate the foreign exchange drain and reduce dependence on extra-national sources.

In 1976 Ghana's internal demand of about 80,000 tons of sugar compared to a domestic production of only 12,000 tons. In 1977 and 1978 production decreased to 10,000 tons and then to virtually zero as foreign exchange shortages limited the import of needed replacements for the state-run industrial sugar complexes.

Promase, Ltd. is one of two small-scale processing facilities now operating in Ghana and producing cane sugar syrup. Each facility has a capacity of 2,000 to 3,400 tons of cane per season. The

technology was adapted from cottage industries of long
experience in many parts of the world. The core of
the process is open-pan evaporation as opposed to the
more industrialized vacuum technology used by major
sugar refiners. Rather than reduce the sugar to crystalline form, these facilities process sugar to a con-
densed liquid form. The product is an acceptable substitute for crystal sugar in most applications in the
baking, confection, dairy products, and alcohol industries.

The open pan process is also compatible with, and
has been employed in, the production of crystalline
sugar products. But the production of sugar syrup is
an alternative offering the advantages of operational
simplicity and lower capital investment.

Product Alternatives

The open pan process has been used in the production of a number of sugarcane products of liquid and
crystalline form. In Ghana, the facilities have produced raw crystalline sugar with molasses byproducts
and high-polarity (pol) and low-polarity (pol) sugar
syrups. Similar processes have been used in other
parts of the world in the production of semi-crystal-
line products (jaggery, panela, gur).

The product alternatives considered for Promase
were:

1. High polarity syrup for external processing
to crystal sugar;

2. Low polarity syrup for direct consumption and
use in sweeteners;

3. Production of raw crystal sugar with molasses
byproducts.

The production of low-polarity syrup offers the
following advantages:

1. It is the simplest of the three processes requiring lower capital investment of cane processed.

2. The process yields about 25 percent (cane
weight) primary product as opposed to 5 to 10 percent
primary product in the case of crystalline sugar.

3. Cane handling for low polarity syrup is

184

greatly simplified. Product quality actually benefits
from delayed processing, and yield is relatively unaf-
fected. In the production of high polarity syrup and
crystalline sugar, rapid and highly coordinated cane
handling is crucial to yields and quality, a requirement difficult to achieve in purchasing cane from
small growers and in areas of limited access and infrastructure.

4. High polarity syrup tends to crystallize in
storage; low polarity syrup is comprised primarily of
invert sugars (glucose and fructose) which do not
crystallize at storage densities. Crystal sugar is
hygroscopic and deteriorates from prolonged storage in
humid environments.

5. The natural acidity of cane extract promotes
the inversion which is implicit in the production of
low polarity syrup. High polarity syrup and crystalline sugar demand the addition of inversion-inhibiting
agents such as lime and close control of the pH
(acidity-alkaline) factor.

6. Even in efficient crystalline sugar plants
the primary product (crystalline sugar) represents
only about half of the total product by weight. The
secondary product (molasses) has limited marketability
as a sweetener. The production of syrup entails no
secondary product. About 25 percent of cane weight is
available as primary product.

In Ghana, a Technoserve food technologist was
able to successfully adapt product formulations to the
use of sugar syrup as a substitute for imported refined crystalline sugar previously used in the baking,
confection, dairy product, and alcohol industries.

Infrastructure

While the Promase concept is not incompatible
with a highly developed infrastructure, neither is it
dependent on it. The plant is located in a rural area
without public water, power, or sanitary services.
Power for cane crushing is provided by a diesel engine.
The plant is self-sufficient in its evaporative heating requirements through the burning of cane residue
(bagasse). The plant has a low-capacity pumped well
driven by a small diesel generator which also provides
power for lighting during night operations.

Although the state-sponsored sugar industry in

Ghana has experienced severe operational impediments related to foreign exchange shortages, Promase, Ltd. has continued to function normally with limited dependence on external supply of replacement parts or raw materials.

Summary

The Promase model is offered as an alternative which may be more compatible, in terms of capital requirements and social and political implications, with the objectives of developing countries.

The conventional wisdom concerning the selection of technology to satisfy demand for processed agricultural products in developing economies has been to opt for centralized, industrialized processes borrowed from the developed countries.

Non-centrifugal sugars (liquid and semi-crylstalline) have been a significant portion of total world consumption and are of growing importance. Liquid sugars are playing an increasingly important part in food processing industries in developed countries.

Questions

1. Why was the open pan sugar production process adopted in preference to vacuum technology?

2. Do its advantages make it appropriate under the circumstances?

3. Explain from the summary: "The Promase model is offered as an alternative which may be more compatible, in terms of capital requirements and social and political implications, with the objectives of developing countries."

Case No. 11

Textile Visual Materials,
Ghana and the Sudan

by
Beverly Emerson Donoghue

This case highlights an innovative education
medium--screenprinted visual aids on fabric--used in
Ghana and the Sudan as appropriate technology. Because
of the great potential of this nonformal medium for
transferring technology in communication, education,
and training in societies with many illiterate and
underprivileged members, it may be significant for development.

Visual materials are important tools for communication, education, and training programs in Africa.
Conventional media and materials, however, are often
scarce because of the reliance on imported supplies,
technologies, and personnel from the developed countries. This paper will examine an innovative educational medium--screenprinted visual aids on cloth--
that was first developed in Ghana.

Introduction to the African Setting

Located on the west coast of Africa, Ghana is
about the size of Oregon and has a population of 11
million. The majority of the population lives in
rural areas--along the coastal plain, in forest regions, and in dry savannah and desert-like terrain.
The primary occupation throughout the country is subsistence farming, although the nation's foreign exchange earnings are mainly from cash crops of cocoa
and timber.

A former British colony, Ghana's official language is English. There are, however, over 55

Abridged and reprinted with the permission of
Beverly Emerson Donoghue, educational materials and
media specialist and graphic designer, with experience
in Africa, Latin America, and low-income Hispanic communities in the United States.

different ethnic groups in the country, each with its own language or dialect. As in most African countries, literacy levels are quite low--about 30 percent in Ghana. Since becoming independent in 1957, one of the government's major tasks has been to expand primary education in order to make it available to all children in the rapidly increasing population. As much as one-fourth or one-third of the annual budget is devoted to expanding and improving the educational system. Extension or nonformal education programs--in agriculture, literacy, health, family planning, community development--have also increased dramatically as government ministries and private organizations have attempted to raise the standard of living of the largely rural, illiterate, and linguistically diverse population.

With such tremendous communications and learning needs, what visual materials, then, are available in Ghana and other African countries? Printed materials are in short supply because, with very few paper production facilities in Africa, almost all paper must be imported and is very expensive. A sheet of paper is truly a luxury in most rural areas.

The situation regarding electric and electronic visual media is also dismal: At the present time [May 1982], all of the equipment must be imported. The cost of the hardware alone is prohibitive in terms of scarce foreign exchange. Software development is limited because of a dependence on imported film supplies and camera equipment. There are few or no spare parts, a shortage of trained personnel for maintaining and repairing the equipment, and an unreliable or non-existent electric supply in rural areas. The most practical projectors available at this time are the small battery-operated filmstrip and slide projectors, costing between $50 and $125 each. These also represent a sizable foreign exchange investment for large-scale programs, but at least they are lightweight, durable, and they work--even in the most remote areas.

The dilemma regarding conventional materials and media can perhaps best be illustrated by three situations in which I was working as a Peace Corps Volunteer for two years. First, as a materials and media specialist at a university, I had access to graphic supplies, photographic facilities and expertise, a slide projector (complete with spare bulbs), and a fairly reliable electric supply. In this instance, it was possible to both develop large illustrations on paper and make color slides from artwork for use in various classes.

At a nurses' training college in the same city, the resources were much more limited. There was only a projector without a bulb (which had been requested over a year earlier), an unreliable electric supply, and no graphic supplies or photographic equipment/facilities. The visual materials "solution" was to paint illustrations for teaching/learning on large plywood sheets for use in the different courses. Although an oddity at first, these life-size murals on wood became very popular with both nursing tutors and their students.

The real nightmare for a materials and media specialist was at the level of the primary schools and extension programs. With such large-scale learning and communication needs, neither printed paper visual aids nor electric/electronic visual media were either technically feasible or affordable--except for the few "Cadillac" programs that received external funding or imported supplies from international aid agencies. The challenge was to look more closely at the available resources, seeking alternative communications media that were practical, affordable, and culturally relevant for both formal and non-formal education programs.

The Rationale for Textile Visual Materials

Unlike paper and electric/electronic visual media, cloth is a familiar and colorful sight in both urban and rural areas of Africa. Everyday scenes of children caring for young ones, fisherfolk bringing in their catch, market women selling their wares--each is accented by colorful cotton prints that are manufactured locally.

While there are no paper production facilities in Ghana, there are several well-established textile factories that weave and print intricate designs in bright colors. Printing facilities vary from large automated factories to small cottage industries.

Throughout the continent, there is a strong tradition of textile artisans, who often create striking fabric designs with the simplest tools. Today village textile industries abound: Weaving, stamping, cassava resist or batik, and tie-dyeing are some of the many

189

techniques used, employing both imported dues and those derived from local plants, trees, and minerals.

Cloth is a very durable material and, unlike paper, will last a long time--despite the temperature and humidity extremes of Africa's wet and dry seasons. Whereas paper is "imported" both from other countries and from the city, cloth is a familiar commodity and as such is much more "touchable" or "approachable" than paper. Cloth can easily be washed when soiled and readily folded up and carried from village to village. People wear fabric, wrap their babies in it, and use it to carry all kinds of things. So why not let cloth carry educational messages as well!

In fact, several African countries have had fabric printed to illustrate and thus promote slogans for national campaigns--such as "Operation Feed Yourself" in Ghana and "Healthful Foods" in Tanzania. When worn, these brightly colored designs become walking posters for everyone to see. The use of printed cloth designs as visual communication tools simply carries the textile medium one step further.

The Production Process

The simplest method for printing large designs on cloth is silkscreen printing. The actual stencil is a very fine mesh screen fabric stretched tightly across a rectangular wooden frame. Open spaces in the screen are the design areas to be printed, with the rest of the screen sealed to prevent ink from passing through.

The basic equipment--screen frame, printing blade, and long printing table--is made mainly from wood. Local materials can satisfy most, if not all, printing supply needs. Most of the labor can be performed by unskilled workers, who can be given on-the-job training in manual screen-printing methods. The labor-intensive printing process is particularly appropriate for developing countries, whose greatest potential resource is the large pool of untrained and under- or unemployed workers.

Research and Development Efforts
with Textile Visual Materials

Because the printed cloth medium seemed to be such a natural one for the African setting, a prototype development project was organized in Ghana in

190

1974 to test the technical feasibility of printing
large educational designs on cloth, to find out how
acceptable the cloth medium would be to educators and
extension personnel, and to determine the production
costs involved. With the assistance of private educational organizations, private industry, and government
agencies, Ghanaian art students designed and printed
on cloth 4-color illustrations of a variety of subjects, including the eye, the digestive system, and a
physical map of Africa. The periodic chart of the
elements and the life cycle of schistosomiasis (bilhar-
zia) were also printed. These 3-foot-by-5-foot illustrations were presented at several national and international conferences and distributed to educators from
various African countries. The response was a unani-
mous preference for this type of cloth-based visual
aid over conventional paper ones. Teachers were delighted to see for the first time educational designs
depicting Africans rather than Europeans--indicating
that the few visual aids used in the schools were imported and irrelevant for the learning needs and African cultural setting. The most frequently-heard comments by teachers and extension workers were feelings
of pride and delight that these vivid illustrations
were produced by Africans, for Africans, and on African soil. Printed cloth designs were seen as one of
the few communications media that could in fact
"reach the village."

As with most printing processes, silkscreen
printing was found to be most feasible economically if
done on a mass-production basis. Based on a minimum
order of 4,000 yards, or 2,400 copies of a 3-foot-by-
5-foot design, preliminary cost estimates made in
Ghana in 1974 indicated that these large 4-color illustrations could be screenprinted on cloth for less
than U.S. $3.00 each. Because cloth was readily
available locally and because the cost of imported
inks and stencil materials was only a tiny fraction
of the overall cost of materials, this production
figure was considered quite reasonable. If the amount
of foreign exchange that would otherwise be spent on
imported paper and imported visual aids was used for
printing inks instead, local industry could produce
more and better visual aids at lower actual cost.

Since 1974, both the economy and political stabi-
lity of Ghana have deteriorated considerably. With
frequent shortages of all kinds of goods, there is an
even more urgent need to develop products from local
materials. Accordingly, an in-depth feasibility study

191

was conducted in Ghana during the Textile Visual Aids Project in 1980, which was jointly sponsored by the Ghanaian government and the U.S. Agency for International Development. The purpose of the study was to determine the availability of necessary supplies; to assess the existing and potential demand for visual materials on cloth; to update production costs; and, if found to be feasible, to recommend organizational options for the development and production of textile visual aids.

Cloth was indeed found to be in short supply. Although cotton-growing is one of the agricultural priorities of the government, most raw cotton fiber must still be imported. Because it would be used for educational purposes, however, an adequate supply could be guaranteed by one or more textile manufacturers. Printing inks were still imported, but some highly successful experiments were conducted, substituting cassava paste for the imported printing binder. It was also found that UNESCO coupons could be used for ordering up to $20,000 worth of imported supplies, more than would be needed to supply a textile production unit with printing ink for 2½ years. For each 2-foot-by-3-foot cloth print used in a school or extension program, the foreign exchange cost would be about 11 cents.

The visual materials needs of various sectors were quite large. There are over 7,000 primary schools and over 4,000 middle schools in Ghana. The Ministry of Health would like to print at least 5,000 copies of designs on several topics. The Home Extension Unit of the Ministry of Agriculture, the Ghana National Family Planning Program, and the Mass Literacy Campaign by the Department of Social Welfare and Community Development are other nationwide extension programs which need long-lasting visual materials. Because of the durability of cloth, agency officials gave $3.66 to $5.50 as a reasonable price range for each textile print.

Production costs were determined for private textile printing factories and for a production unit to be based at a specialist training college with printing facilities. With a minimum order of 1,000 prints, the cost per print from a private textile firm would be $2.28. If done by the production unit, the cost would vary from $1.86 to $2.74 each, depending upon the quality of cloth used for printing. This is well below the acceptable price range.

A specialist training college in a small town
hosted a Textile Visual Aids Workshop, providing facilities and a counterpart/production manager. A production unit there would offer valuable training to the
textile students while producing not only attractive
but useful textile designs. In addition, channeling
resources out of the capital and other large cities
and into a rural institution not only makes use of existing facilities but supports the stated government
policy of encouraging rural development.

The 10-week workshop on Textile Visual Aids was
provided for representatives from various ministries
and organizations associated with nonformal education
and family planning. In addition to experimenting
with the cassava-based ink, the group also began experiments with local dyes for substitution for imported dyes and pigments. Although not as washable as
the imported ones, the local dyes deserve systematic
research.

The designs developed by workshop participants
included the following topics: Raising rabbits for
food, eating a balanced diet, making oral rehydration
fluid, preventing diarrhea, family planning (one directed at women, another directed at men), and village
scenes for language learning in primary schools. The
various ministries were quite pleased both by the
quality of designs developed and by the use of cloth
as a communications medium. Because several ministries have overlapping needs, it was recommended that
the textile visual materials production unit as proposed be a collaborative effort of interested institutions.

In the Sudan, the development of textile visual
aids is even more advanced. The abundance and low
cost of cotton there make it a very appropriate visual
communications tool. All of the materials needed are
locally available, including the following innovative
adaptations: Gelatin glue for preparing the screen
stencils; women's veil material for the screen fabric;
and sorghum starch and direct dyestuffs to make the
dyepaste/ink.

Plans are now under way for the World Health Organization and the Ministry of Health to establish a
rural production center in the Sudan for printing textile visual messages for health education programs.
It is possible that, once established, the production
center may evolve into an income-generating operation

193

for the village--printing cloth designs for the many extension programs in the country.

Textile visual materials are not offered as a panacea for the urgent communication and educational needs in Africa. Visual aids are clearly very helpful, particularly when combined with radio discussion groups and other participatory media. What is significant, though, is the approach: Rather than trying to transplant a communications medium from the West, the strategy has been to take advantage of the materials and resources that are available locally--to experiment--so that communications tools will be relevant for the local resources, learning needs, and cultural setting in African countries.

Questions

1. In the Ghanaian setting, why was it necessary to seek nonformal education programs? Are the conditions leading to this necessity typical of developing countries?

2. Why is "the labor-intensive printing process... particularly appropriate for developing countries"?

3. Why is printed cloth socially and culturally relevant compared to other visual aids?

Case No. 12

Tetanus Vaccination

by
Medicus Mundi Internationalis

This case highlights the appropriateness of the transfer of preventive and curative medical technology from developed to developing countries to fight a serious pathological condition in part of the Third World.

In some regions of West Africa 14 percent of the newborn die before their 10th day of life because of tetanus of the umbilical cord. Every hospital has to treat such cases. They require permanent supervision by a physician and if nurses work in eight-hour shifts, three nurses round the clock are tied up on one case. The physician may be content if by using all means of modern treatment he is able to save four cases out of 10. This is curative medicine.

It may make sense in a region where cases are frequent to immunize all mothers during pregnancy against tetanus. This requires three injections. Babies of vaccinated mothers are practically exempt of cord tetanus. This is preventive medicine; or better still, it is prevention because there is no patient to be cured.

But in Western Europe cord tetanus is practically extinct. Why? Europe is clean and people are well educated. Everybody has tap water and clean sheets can be put on every bed. There are many trained midwives and transport is easily available. The usual cause of cord tetanus is a dirty bamboo knife with which the cord (traditionally) has to be severed or if the cord is contaminated with earth.

If the mother is delivered on a clean bed, if the cord is cut with boiled scissors, the problem of cord tetanus is abolished. This is domestic science and in this respect it is achieved by a process of

---

Reprinted from Concepts I with the permission of Medicus Mundi Internationalis.

health education. Even in its cheapest varieties it is of the utmost importance for health promotion. It can be practiced without any doctor or midwife.

Although it is possible in theory to distinguish between these different approaches to the advancement of health, health in reality is indivisible. In practice, its curative, preventive, and promotive aspects are closely correlated. The more defective the health situation in a given community, the more axiomatic is a three-pronged attack.

Questions

1. Comment on the paradox that Western technology able to prevent umbilical cord tetanus is now more appropriate in developing countries than in the developed countries from where it originates.

2. Which human and social conditions make it so?

3. What would be the ideal form of technology transfer to combat umbilical cord tetanus? Why is it not practiced?

Case No. 13

The Baby Killer: Formula Milk

by
Medicus Mundi Internationalis

> This case involves the transfer of powdered baby
> milk formula. It shows how a technology that played a
> major role in reducing infant mortality in the West in
> the 19th century is still completely inappropriate in
> the different setting of developing countries today.

Third World babies are dying because their
mothers bottle-feed them with Western style infant
milk. Many that do not die are drawn into a vicious
cycle of malnutrition and disease that will leave them
physically and intellectually stunted for life.

Mother's milk is accepted by all to be the best
food for any baby under six months old. Although even
the baby food industry agrees that this is correct,
more and more Third World mothers are turning to artificial foods during the first months of their babies'
lives. In the squalor and poverty of the new cities
of Africa, Asia, and Latin America, the decision is
often fatal.

The baby food industry stands accused of promoting its products in communities which cannot use them
properly; of using promotional sales girls dressed in
nurses' uniforms give away samples and free gift gim-
micks that persuade mothers to give up breast feeding.

Where there is only squalor, the choice of an
artificial substitute for breast milk is in reality a
choice between health and disease.

The results can be seen in the clinics and hospitals, the slums and the graveyards of the Third World.
Children have bodies that waste away until all that
there is left is a big head on top of the shrivelled
body of an old man. These are children with the ob-
scene bloated belly of kwashiorkor.

Reprinted from Concepts II with the permission of
Medicus Mundi Internationalis.

197

Why are mothers abandoning breast feeding in countries where it is part of the culture? Are we helping to promote the trend? What is the responsibility of the baby food industry? What are we doing to prevent avoidable malnutrition?

Questions

1. What makes baby milk formula inappropriate in a Third World setting?

2. What would it take to promote the use of mother's milk as energetically as what was done with formula milk?

3. How would you answer the four questions in the last paragraph of this case study?

Chapter 4
Problems in Technology Transfer

These cases deal with the problems in selecting and assessing technology transfer. Notice how the occasional paucity of information regarding all the parameters of a given technology transfer, lack of sociocultural sensitivity, and the need to choose between trade-offs may account for ostensibly unsuitable selections and incorrect assessments.

Case No. 14

Solar Energy Devices, Lesotho

by
Roy Lock,
Appropriate Technology International,
Washington, D.C.

This case highlights some technical, economic,
social, and other constraints in transferring technology alien to and with little participation by the recipient community. The case also underscores the importance of assessment after a transfer is made.

I. Overview of A.T. International Involvement

Description of the Thaba Tseka Solar Energy Project

Phase I of this project provided for two American
experts in renewable energy to work in the mountainous
region of Lesotho with the Thaba Tseka Integrated
Rural Development Project (TTRDP) in order to:

● Identify practical solar energy devices suited
  to the cultural, economic, and ecological climate
  of the Thaba Tseka District to meet domestic,
  agricultural, and institutional energy needs;

● Train local people to build, install, and maintain these devices; and

● Help establish small businesses, producing and
  selling solar energy devices, to increase local
  employment and economic self-sufficiency.

On completion of this phase, in May 1980, the consultants prepared an exhaustive report of their
achievements describing the failures as well as the
successes and attempting to evaluate the impact of the
work. Among the conclusions was a strong recommendation that the work be continued for a further period,
employing only one of the expatriate consultants.

Abridged and reprinted with the permission of
A. T. International.

This recommendation was endorsed by the TTRDP management, and A.T. International (ATI) agreed to further funding of a second phase over a seven-month period.

This second phase, it was claimed, would permit continued monitoring and wider dissemination of technologies introduced in Phase I, and would specifically result in:

- A selection of fully tested solar devices for cooking, water heating, food drying, and battery charging with details of costs, markets, and most suitable manufacturing options, many of which will have been made by villagers themselves or sold to them;

- Demonstrated improvements--outlining costs and benefits--which can be made to the local houses (rondavels);

- Improved and installed wood and dung-burning stoves and experience in the costing, marketing demand, and methods of manufacture; and

- Four trained Basothos (natives of Lesotho) with expertise in extension work on appropriate technologies for villages, and with a special understanding of the principles and manufacturing techniques applicable to solar devices and cooling methods.

By the end of December 1980, this phase, and therefore the ATI-funded project, was completed. A final descriptive and evaluative report was produced by the consultant and was widely distributed among interested groups throughout the world. The Basothos who were trained have successfully continued as a unit of the Rural Technology Unit (RTU) of the Thaba Tseka Integrated Rural Development Project and the dissemination of the technologies introduced, as well as some research and development work, is continuing. The American consultant employed has been hired by the USAID/Ministry of Rural Development (of the government of Lesotho) Renewable Energy Technology (USAID/RET) Project and has presumably brought to that project both his expertise and the results of his findings at Thaba Tseka.

Rationale for ATI's Funding and
Achievement of Objectives

Without the benefit of evidence from a proper evaluation of this project, ATI staff were satisfied

202

that many of its (ATI's) objectives in funding the project were achieved. This opinion prevailed despite
several obvious areas in which the project itself fell
short of its own stated objectives. This point bears
closer examination.

In agreeing to support TTRDP and fund the Solar
Energy Project, ATI was aware that several factors
were at odds with the then prevailing philosophy at
ATI.

● TTRDP was almost exclusively managed by expatriates
  and largely funded by overseas donors.

● The original project idea was that of expatriates.

● The "project" consisted largely of expatriates bringing in their ideas of what was "appropriate" and
  offering it to the local people.

● Participation of local staff was promised, but was
  not in evidence, at the start of the project.

● The desire, as well as the necessary financing, to
  continue the work beyond the ATI funding period
  were far from evident.

ATI staff considered these points very carefully. It
was and is still a valid question to ask what a donor
agency's attitude to such a situation should be (i.e.,
where there is little or no interest in the technology
by the local population; where there is no infrastructure run by local people; and where the technology is
experimental and therefore makes long-term planning of
its development hypothetical, but where nonetheless,
through "Western eyes," it seems like an excellent
idea).

ATI staff decided to give more weight to other
considerations and were thus able to answer their own
question simply by implying that a donor agency should
do something, rather than nothing. These other considerations were as follows:

● A genuine desire by ATI to create a long-term program in Lesotho and to become a major force in the
  region in the support of the appropriate technology
  (A.T.) movement. This program initially took the
  form of supporting the three prominent appropriate
  technology activities in the country at that time--
  Thaba Tseka, Thaba Khupa, and the national A.T. unit

at Basotho Enterprises Development Company (BEDCO, a parastatal).

● A desire by ATI to develop a lasting relationship with TTRDP and to become influential in determining its management structure as well as its specific activities. TTRDP was being looked upon by the government of Lesotho as a possible model for the eventual complete decentralization of the government.

● An expressed serious misgiving on the part of several well-placed development officials concerning the USAID/RET project that was then being planned. During the various phases of the planning for this multi-million dollar project, concerns were voiced to ATI staff about its design: Thaba Tseka and Thaba Khupa were not to be involved at all; all technologies were "foreign" and to be introduced by expatriates; the project was too big, too ambiguous, and too expensive; and beyond a minimum delay of at least one year, no one had any idea when it would ever start. A desire by ATI, therefore, to influence the eventual character of the USAID project was an additional factor.

In terms of TTRDP's stated objectives for the Solar Energy Project, the results of one year's effort show many areas in which it might reasonably be said to have failed in these objectives. A cursory glance at the appendix, produced at the end of Phase I of the project, shows that in absolute terms very few devices were sold at all. The position at the end of Phase II shows little improvement upon that except that 10 metal ovens had been made and sold, and that the improved paola (a non-solar cooking device) was in regular production at the RTU and was, and continues to be, selling well. No more photovoltaic battery chargers have been installed at clinics, no food dehydrators are being used on a commercial scale by farmers, and no major changes have been observed in rondavel design. Especially disappointing from ATI's viewpoint is the fact that very little success appears to have been achieved in the commercialization of the devices. No local craftsmen, either at Thaba Tseka or in the capital of Maseru, appear to be interested in manufacturing and selling the devices for their own profit. Only the RTU has established a couple of devices which it can construct and sell to local people successfully.

These results are disappointing, but they are not

204

wholly unexpected. Indeed, it would have been more surprising if a whole range of solar devices had become suddenly very popular and in great demand in the Thaba Tseka region. "Development" rarely works that way.

The project must be put into perspective. Two expatriates worked for six months, and one expatriate for a further six months, on a range of over 25 separate devices. They were supported by no more than four untrained local people and for much of that time by fewer than four. They worked in a region where transportation and communication are extremely difficult. And, they were introducing a completely new concept to uneducated, poor people.

What was produced, on the other hand, can be considered impressive in view of the above constraints. Over 25 separate devices were built, tested, and to some extent disseminated. The dissemination was neither wide nor in great depth, and the results are therefore inconclusive for most of the devices. Nevertheless, the project produced the following:

● Full technical documentation on the devices built and tested, including costs.

● Some indication of local acceptance in cases of devices which were totally rejected and those in which some interest was shown.

● Four trained Basothos with experience in construction and dissemination of these devices, who are continuing to operate as a unit within the RTU.

● Operating models of many of the devices, which are seen constantly by the local population.

In terms of ATI's perspective, the following was achieved by the project:

● Thaba Tseka's credibility in the field of appropriate technology has risen to the level where it is now considered the leader in the field in Lesotho.

● Thaba Tseka is now prominent as a development area for the USAID/RET project.

● A body of relevant data as well as considerable interest in solar energy existed in Lesotho at the start of the USAID/RET project.

205

• A senior expatriate manager of the USAID/RET project brought with him at the start of the project over a year's experience in Lesotho working on the Solar Energy Project at Thaba Tseka.

Finally, it must be pointed out that in considering the following discussion of how this project supports the stated theses, no formal data base evaluation of the project has so far been undertaken to provide useful information for this paper. An appropriate time for such an evaluation would probably not be before the completion of the USAID/RET project, the success of which will no doubt be influenced by the effectiveness of this Thaba Tseka solar project. The following discussion, then, is based largely upon very limited information and a considerable amount of conjecture.

II. The Project Context

Nature of the Sponsoring Organization
Lesotho and the Thaba Tseka Region

The Kingdom of Lesotho is a small mountainous country of about 30,000 square kilometers, land-locked, and completely surrounded by the Republic of South Africa. Its population of some 1.2 million live mainly in the western part of the country, known as the lowlands; the eastern part, comprising perhaps two-thirds of the land area, is mountainous, rugged, and largely inaccessible except on horseback. Nonetheless, a large population lives in this harsh area, maintaining a subsistence agriculture of wheat, maize, sorghum, sheep and goats, and some cattle. Every family owns horses, which are almost exclusively used for transportation and only very rarely for traction in agriculture. The subsistence economy is supported with rather dramatic effects by income from over 50 percent of the male labor force who are under contract to the mines in South Africa and thus out of the country for periods of up to two years at a time.

For villagers living in this region, the prospects for improving their lives are poor: The soil is constantly eroding, the climate is harsh with long periods below freezing in winter, and there are almost no trees. The common fuel for cooking and heating is dung, which burns very well, but its use depletes the soil and causes health problems. Imported fuels are used, but are expensive and supply is difficult and undependable. The sun, however, is plentiful, especially in winter when the need is greatest. Technically, the region would appear to be ideally suited to exploit the possibilities of solar energy.

One further point worth mentioning here is that the migrant-worker aspect of family life complicates the notion of what is culturally and socially acceptable to the people of the region. Many of the families have access to relatively large amounts of cash when husbands or sons return from the mines. This is often mismanaged. In addition, access to sophisticated technological products by members of the family working in the Republic of South Africa and the close proximity of the Republic to any point in this small country strongly influence the acceptance of what might be described as "second rate" appropriate technologies among large sectors of the population. Easy access to these "sophisticated technologies," together with occasional cash with which to acquire them, has led to aspirations beyond the real economic reach of the people.

ATI Involvement with TTRDP

ATI staff first visited Thaba Tseka in the summer of 1978. Having established a funding relationship with the National Appropriate Technology Centre in Maseru, Thaba Tseka presented a project, some of whose goals at that time were in line with those of ATI. Projects concerned with appropriate technology in Lesotho at that time were few. ATI established a working relationship with each of them and, together with its link with the National A.T. Centre, had visions of a countrywide program of assistance in Lesotho.

At the end of 1978, an American energy consultant, Alan Wyatt, arrived at Thaba Tseka. He was traveling through Africa, stopped at Thaba Tseka, and showed an interest in the work being done there. The project managers were interested in him and hired Wyatt, from project funds, to study the potential for wind and solar energy usage. His work involved the setting up of measuring devices in the region, and early conclusions were that wind energy showed less immediate potential and required greater study than the more obvious benefits of solar energy. Responding to demands (for example, from a local clinic), Wyatt built a successful windmill to generate electric power. The demands from institutions, however, were for hot water, cooking facilities, distilled water (for clinics), as well as electricity for lighting. Attention on renewable energy therefore naturally focused more sharply upon solar energy potential.

In the spring of 1979, Wyatt designed a project to work on solar energy for six months. He proposed that this be undertaken by himself and an American with whom he had worked previously. In April 1979 this proposal was presented to ATI by the TTRDP project director and the head of the technical division. ATI's concern at this point was with the commercialization of the technologies, and this aspect was included in the project design. After interviews with Wyatt and the other proposed consultant, Gary Klein, ATI agreed in August 1979 to fund the project.

By May 1980, the period of the project was completed and a final report and evaluation were produced by the two consultants. This report showed some positive and some negative results. Its major conclusion was a recommendation that work should be continued for six months. The arguments for this, it was claimed, included: Possible dissemination of the devices already tested, the need for some modification and continued testing through the winter months, promotion and test modification of the traditional Basotho house, and a need to concentrate on improved non-solar cooking methods. The proposal was for Klein alone to undertake this work, Wyatt by this time having decided to return to the United States for personal reasons. Also specifically stated in the proposal was a plan to train four Basothos in the theory of the devices and in the art of disseminating the technology.

ATI agreed to fund this second phase of the project, which was undertaken between May and December 1980. Early in 1981, the project having been completed, the consultant produced a second report, equally useful in terms of technical and evaluation data.

ATI staff have visited Thaba Tseka 14 times during its almost four-year involvement with TTRDP. On most of these occasions, ATI has been approached to fund additional activities within the project. Many of these have been considered by ATI staff in some detail.

In the context of responding to these requests, ATI regularly discussed its concern that TTRDP had not yet clearly defined its organizational structure.

Questions of management capability and counterpart training possibilities for these proposed activities were raised. Although ATI did not play any role beyond these discussions in assisting TTRDP to resolve such issues, it did make it clear that they were the reason behind its refusal to fund additional activities at that time.

The State of the Art of Technology

Technology Overview:
The Lesotho Experience

To the people of the mountainous Thaba Tseka District of Lesotho, the concept of solar energy was, in many respects, new. Admittedly, it has been customary to dry meat and vegetables in the sun by leaving them uncovered on the ground as long as necessary in order to store the dried foods beyond the growing season (food storage is a major problem of the area). But, apart from this, harnessing or focusing the power of the sun was not done before the start of the ATI-funded project.

In the capital city, Maseru, domestic solar water heaters have been in use for some years. A cooperative housing project, for example, installed them as standard throughout a small estate which was built. Most of these systems, which basically consisted of two panels and an inside tank, were manufactured in and imported from the Republic of South Africa.

Some work had been done at the Ministry of Rural Development on parabolic solar cookers, food dehydrators, and improved solid fuel cookers. But the work was experimental, the results confusing, the technology not extensively tested and not at all disseminated, and the products prohibitively expensive. This had been carried out by a Swiss expatriate, but by early 1979 all work had ceased.

In early 1979, therefore, the Thaba Tseka project had no solar energy technology experience and no plans for any work in the area.

Choice of Technology:
Original Project Objectives

Wyatt's proposal was for work on the following devices: Solar stills, solar cookers, solar water heaters, solar battery chargers, batch (institutional)

209

water heaters, solar food dehydrators, greenhouses (private and institutional). The proposal also included two weeks' work in studying the local housing style and (thermal) improvements which could be made to it.

ATI formally agreed to fund the project and Wyatt and Klein, the second consultant, toured the United States and Canada for several weeks visiting institutions involved in solar energy work.

The consultants visited Botswana and several organizations in the Republic of South Africa on their way back to Lesotho to begin the project. Their first move on the project was to visit, together with three of the four Basothos hired to work with them on the project, six villages in the Thaba Tseka District. This "survey" lasted 10 days, and pitsos ("village meetings") were held in each village to try to determine the needs and priorities of the villagers.

Changing Priorities:
The Feedback Process

The pitso is a meeting of all the villagers, semiformalized by the attendance of the village headman. During the course of the project, the RTU staff developed a format for these meetings and were successful in eliciting the opinions of the villagers. At first, the RTU staff led the meetings throughout by asking questions. At later meetings, when various devices were to be demonstrated, the relationship changed. The villagers tended to lead the meetings by continuously asking questions of the project staff.

The project staff informally assessed the interest of the villages by the questions asked and (therefore) the apparent interest shown in each device. Initially, this reaction determined to some extent where the project staff placed the priorities of their work. In some cases (the parabolic cooker, for example), this method of determining the needs of the villagers proved misleading. In this instance, the villagers were fascinated by the "spectacular" results of the demonstration, but, perhaps because the results were so spectacular and ill-understood, there proved to be almost no interest in actually acquiring the device.

The more useful feedback was revealed several months later. Some devices were left in the villages

210

after the pitsos, on loan to the villagers. Project
staff revisited the village several months later and
inquired about the use made of the devices. They
found this to be a more accurate way of gauging local
interest. Some devices had not been used at all,
whereas villagers were anxious to purchase others.

The overall dissemination process was multi-
faceted:

● The pitsos and village construction workshops were
held throughout. It is estimated by the project
managers that approximately 12 percent of all households in the district had some member of the household attend a demonstration. Two percent of those
who attended actually purchased a solar device.

● Much of the construction and testing work was carried out at the RTU workshop at Thaba Tseka village.
The workshop was visited by many villagers from the
immediate vicinity of TTRDP.

● Short-training workshops were held at the RTU where
village craftsmen were taught how to build certain
devices. But no cases were recorded of such craftsmen employing these additional skills and actually
selling any device to a customer.

Overall Conclusions

"If we were to do this project again, what would
we do better? What have we learned?" The answers
given below are the views of the writer. The consultants who managed the projects have not been asked for
their views.

● Involvement of the end-user was critical, though admittedly difficult in terms of operating technique.
It was only from this source that the project managers learned of what was likely to be successful.

● The people of the Thaba Tseka region were most interested in devices which were both field-proven and
which worked. Devices which developed flaws or
which did not perform under certain circumstances
were rejected.

● The devices which were simple and built from locally
available materials were more readily accepted. The
"magical" devices were found to be fascinating, but
little in demand. Most popular of all were those

211

which were adapted from traditional technologies.

APPENDIX

Analysis of Devices Produced and
Testing During Phase I

	Built by	Number Built	Number Sold	Approx. Unit Cost (U.S. $)	Potential Market**
1) Stone cold frames	Trainees	2*	0	25	M
2) Solar cooker oven	Consultants & trainees	1	0	40	M
3) Solar cooker parabolic oven	Trainees & RTU	1*	0	50	?
4) Solar cooker-steam cooker	Consultants & trainees	2	0	75	L
5) Solar electric system clinic radio & lighting	Consultants & trainees	1	0	1,770	5-30 clinics
6) Hay box-grass & cardboard	Trainees	3	1	2	M
7) Hay box-styrofoam pillows & cardboard	Trainees	1	0	5	?
8) Water heater-village sheet metal	Village handyman at RTU	10	2	12	M
9) Water heater-village wood	Village handyman at RTU	5	1	8	L
10) Water heater-village stone & mud	Trainees & villagers	15	10	4	H
11) Water heater-commercial Van Leer	South Africa	N/A	0	230	L
12) Water heater-commercial BEDCO	Private company (BEDCO)	N/A	0	115	L

212

APPENDIX CONTINUED

	Built by	Number Built	Number Sold	Approx. Unit Cost (U.S. $)	Potential Market**
13) Food dryers-village	Private company (Minrudev)	N/A	0	35	L
14) Food dryers-village-modified Minrudev	RTU & trainees	1	1	40	L
15) Food dryers-village stone & mud	Trainees & villagers	3	2	20	M/H
16) Solar green-house (large)-attached to house	Trainees & consultants	1	0	770	M
17) Solar green-house (small)-attached to house	Trainees	1	0	40	M/H
18) Solar rondavel	Trainees & consultants	2	0	230	H

* Under construction

** Market potential based on end-user interest as indicated by number of positive responses from villagers:

  L = Low: fewer than 50
  M = Medium: fewer than 100
  H = High: more than 100

Note:
The estimates of potential market categorized above and applied to each device appear to be purely guesswork: It is interesting to note that those devices marked (H) --stove and mud water heaters, stove and mud food dehydrators, small solar greenhouses, solar rondavel--were not especially exploited during Phase II and did not achieve anywhere near the market potential envisioned. From this phase, the mud and stove water heater proved the most successful in Phase II, but was overshadowed in success by the sheet-metal water heater, solar oven, and the paola.

213

Questions

1. What is the impact of large, intermittent cash
flows from migrant labor in South Africa on attitudes
toward technology in Lesotho?

2. What method of assessment was used to evaluate the
effectiveness of solar devices? Was this method con-
clusive?

3. Explain this quote from context: "...the
villagers were fascinated by the 'spectacular' results
of the demonstration, but, perhaps because the results
were so spectacular and ill-understood, there proved
to be almost no interest in actually acquiring the device" (the parabolic cooker).

Case No. 15

Two-Way Radio for Rural Health Care

by
Douglas Goldschmidt, Heather E. Hudson,
and Wilma Lynn,
Academy for Educational Development,
New York

This case highlights the selection of an alternative and appropriate technology for rural health care and other purposes--two-way radio systems--in the absence of an operational commercial telephone network or other means of communications.

Introduction

Difficulties in Rural Health Care Delivery

The growing emphasis on the needs of the rural poor in the developing world has focused attention on creating innovative means of extending health care to rural areas. It has been evident for some time that it is difficult and expensive to extend physician-based health care into rural areas. Not only are there shortages of physicians available and willing to serve in such areas, but the use of physicians in such contexts is often a poor allocation of resources. Rural areas usually lack the supporting medical infrastructure which physicians rely on--laboratories, diagnostic equipment, hospitals, and the like. Further, the provision of such an infrastructure would be prohibitively expensive. However, without an alternative to physician-based care, rural populations are left without any medical care.

The innovations of the Chinese "barefoot doctor" system, and of similar efforts to train cadres of rural health workers in other countries, have modified the approach to rural health care. Rather than

Abridged and reprinted from Douglas Goldschmidt, Heather E. Hudson, and Wilma Lynn, "Two-Way Radio for Rural Health Care," with the permission of the Academy for Educational Development. Report originally prepared for the U.S. Agency for International Development.

relying on physicians, health planners for populations are turning to paraprofessionals at varying levels of training. These paraprofessionals, consisting of nurses, health aides, midwives, and medics, provide "front line" rural primary and preventive health care.

However, the very isolation of rural areas creates difficulties for provision of even these relatively simple medical services. Ordering drugs and supplies can take several weeks or months--usually far too long to wait, particularly during epidemics. The transmission of data for laboratory tests and the results may take so long at times that test results are of interest only to the statisticians. Emergency referrals to hospitals may wait for days in many areas for transportation to be arranged. In treating complicated cases, health workers must rely on their own limited training and skills, or risk an often difficult and expensive evacuation.

Aside from the various urgencies of health care, routine administrative and medical procedures may be significantly delayed, or prevented, in the absence of reliable communications. Further, for many of the medical personnel, the isolation of the rural communities can contribute to loneliness, which in turn induces higher levels of staff turnover and, most importantly, stagnation of skills. Without regular interchanges of information on medical procedures, the field staff can quickly fall behind in developing new skills, or even in maintaining current knowledge.

In such an environment, the possibilities of two-way telecommunications have become attractive. A two-way communications system offers the possibility of regular administrative and medical consultations, as well as a practical medium for other official and non-official communications.

Two-Way Communication Technologies

Several technologies may be used for two-way communication. All permit two-way voice communication between two sites; some can be used for conferencing among several sites. They vary in cost, range (distance), quality of signal, and reliability.

Telephone communications are generally the highest quality two-way communications available. Assuming that service is maintained at a commercial level, telephone networks should provide 24-hour reliable

216

communications of high audio quality. Unfortunately, the telephone networks are extremely expensive to construct and are only beginning to be extended into rural areas. Further, many of the telephone networks of LDCs are of marginal quality in general and of submarginal quality in the rural areas.

In areas where commercial telephone service has not been provided, a variety of two-way radio systems is used. High frequency radio is perhaps the most common. High frequency (HF) communication works by bouncing radio waves off the ionosphere. HF communications can cover great distances of hundreds or thousands of miles but with varying reliability. However, in regions where distances exceed line of sight between locations (usually 50 miles or less), HF offers the most practical means of communications.

Very high frequency (VHF) radio can be used where distances between communities are less than 50 miles or where a hill or mountain can be used for a repeater that allows the signal's line of sight to cover a much greater area. The reliability of VHF communication is generally good within its limited range.

Citizens' band (CB) radios offer very inexpensive communication but over short distances of about 5 to 20 miles. CB radios are small and portable, but are generally not ruggedly designed for rural field settings.

Two-way radio systems most nearly fit into the category of appropriate technology for rural health care. In the absence of other telecommunications systems, two-way radio systems, in a number of different forms to be discussed below, can provide varying degrees of reliability at a low capital cost. These systems, given present technology, are easy to maintain (although not always maintained in practice), easy to use by field personnel, and can provide years of service pending the eventual installation of regular telecommunications services.

This report is concerned primarily with experiences with two-way radio systems, as these are currently the most likely technology to be used in future rural health programs. This should not imply, however, an uncritical endorsement of two-way radio systems. Such systems are almost invariably inferior to a properly functioning commercial telephone system in terms of reliability and, in the long run, cost (both capital and operational). Also, the proliferation of two-way radio systems can lead to unnecessary frequency congestion and can actually impede the development of regular telecommunications services.

Functions of Two-Way Communication

Medical Consultation and Referrals

It is generally assumed that the most critical use of two-way communications for rural health is connecting the rural health practitioners with physicians and nurses in regional or national hospitals. Such contacts are for consulting about a patient's condition for both diagnostic and prescriptive advice, determining whether a patient should be referred to a hospital for treatment, and following up on the condition of a patient either at the hospital or the field location.

The need and importance of this type of communications depend heavily on the medical protocols of the particular project. In Alaska, the Public Health Service holds regular "doctor calls" for the rural health aides--when the doctor contacts each health aide daily to provide consultation and handle administrative matters. These aides have minimal training (some less than three months) and apparently require skilled outside interventions.

In Guyana, the medex (health workers trained by the MEDEX project in Guyana) are generally expected not to consult with the physicians by radio except in the case of emergencies requiring referrals to hospitals, or in very difficult diagnoses. They have developed an extensive handbook of medical protocols which indicates what steps they should take in various situations, including emergencies. The MEDEX field staff are not encouraged to call MEDEX headquarters for consultation except in cases falling outside the protocols. These medex have more than one year of training and are expected to be self-sufficient.

The existence of a two-way radio system allows varying degrees of consultation between the field and professionals depending on the reliability of the radio system and the availability of professional staff. Improvements in radio reliability generally increase communication costs. A radio network which operates for 24 hours a day is more expensive in terms of number of frequencies and monitoring time required than a simple HF system which can operate only during parts of the day. So, the necessity for medical consultations during the evening must be evaluated in planning the radio system.

Further, utilizing professional staff as radio consultants requires that the staff be available for such consultations. Professional time requirements will vary with the number of radio sites and with the dependence of the field staff on professional backup.

Administration

Two-way communications play a critical role in the administration of health programs. This is particularly the case in ordering drugs and medical supplies. In remote locations, epidemics or less serious outbreaks of disease may require immediate shipments of antibiotics, vaccines, and the like. However, given the rural isolation, with infrequent transportation and thus slow mail service, it may take days, or weeks, for news of these needs to reach headquarters. Further, as AID consultants observed in Guyana, the drug order may be received and headquarters may believe that it has been shipped, while in fact the drugs remain in some warehouse, ignored by the shipping agent. The radio system can then be used to monitor the drug shipments.

Similarly, the radios can be essential for ordering and shipping routine supplies such as food, furniture, fuel, and spare parts. As with medical supplies, these items are often needed quickly, and quick delivery may not be possible to arrange without rapid communications. Also, the existence of the two-way system allows headquarters to determine precisely what is needed if an order is vague or if the requested supplies are not immediately available.

The radio can also be used for routine administration, such as arranging for vacation replacements for field workers, arranging for messages to be relayed to various people at headquarters or in the field, and the like. While many of these functions do not have the same time value as ordering of drugs and supplies, the existence of the radio may not only speed these processes along, but may also significantly improve field morale.

Coordination of Transportation

One of the greatest difficulties in rural medical care is arranging transportation for critically ill patients from the field to a regional or national medical center. Rural areas often lack regular or frequent transportation. As a result, during emergencies special transportation must be arranged. One of the more innovative responses to this problem has been AMREF's (African Medical and Relief Foundation, Nairobi, Kenya) Flying Doctor Service in East Africa. This service ties field professionals to each other, to hospitals, and to AMREF headquarters via a two-way radio system. The radios may be used to alert the doctors who can either fly to the site of the emergency or arrange for one of AMREF's airplanes to evacuate the patient. Generally, they try to move patients to clinics and regional hospitals where specialists can be flown in. The close ties with AMREF between communications and transportation allow for a flexible and rapid response to various types of medical needs.

Other rural health systems use radios to arrange emergency evacuations. One of the most extensive two-way communications systems, the Public Health Service's satellite communications network in Alaska, has been used heavily for arranging air evacuations in the absence of any road transportation in the Arctic. While evacuations are costly, they can be effective in saving lives if combined with a communications system which can maximize the speed of evacuation. Such unions of transportation and communications are particularly effective when, as in Alaska, East Africa, Guyana, and other locations, both the medical and transportation systems have communications systems which allow messages to be conveyed rapidly from one to the other. For example, AMREF planes have one of the terrestrial network's frequencies. In Lesotho, the Flying Doctor Service wanted the radio system designed to include a network frequency in its plane.

In addition, the evaluation of the Alaska satellite network showed that some evacuations could be prevented if expert advice were available by radio.

Continuing Education

The isolation of rural health workers makes continuing their training or even providing refresher courses very difficult. Although there have been efforts to enable field workers to return for refresher

220

or training courses regularly (generally once a year), it is difficult to provide in such courses the range of problems a worker might encounter during the year.

The level of education courses can vary tremendously according to the time devoted to preparing materials and to studying in the field. The simplest types of training occur through field workers discussing their problems during consultations in a conference-call situation. This type of training is used extensively in Alaska, based on experiences during the ATS-1 satellite demonstration. Under this procedure, health workers in a particular area are on alert at the daily doctor call during which each aide describes cases and the doctor provides diagnostic and prescriptive assistance. Listening in allows each aide to hear varying descriptions of illnesses, learn ways of describing symptoms, and become aware of diagnoses and treatments possible for various symptoms.

A more directed approach has been adopted in Guyana. Once a week a general conference is held during which the physician presents a case which has been referred to Georgetown during the preceding week. The physician presents certain symptoms and quizzes the medex on them, the types of diseases associated with the symptoms, approaches to diagnosis, and the like. This session acts as a major review of particular illnesses and treatments and has been highly recommended by both physicians and field workers.

An even more sophisticated system utilizes formal continuing education programs based on interaction between the headquarters and the medex. Such a program has been developed by Judy Roberts as an outgrowth of the HERMES (CTS) satellite program in Canada. This form of education presents new ways of assessing medical problems through a series of presentations, questions, and follow-ups over a period of time. While such a training program would appear to have significant promise for long-range planning, it also has some major difficulties, primarily in terms of the time required. Such a program requires a major effort by a physician, or some other medical expert, in developing the curriculum. Also, the health workers must have time to participate in the program during the hours the radio system is operational.

The limitations of HF radio systems in terms of reliability and signal quality make this kind of intensive training difficult.

221

"Bush Telephone" Uses

It is probable that the health system's radio network will be used for more than simple health communications. In the absence of a telephone system, rural radio networks are used for a broad range of communications. Most commonly, messages are relayed either to headquarters or to another field post for passage to the recipient. The intensity with which the health system is used for this purpose generally depends on the health requirements of the system and the goodwill of the professionals. However, there may be increasing pressure to use the system for non-health purposes in the absence of other public telecommunications systems or otherwise available radio systems.

The first non-health users of the system will generally be the field practitioners, such as extension workers or teachers, who will use the radio to send personal messages and simply to communicate with other field workers. These are important functions as they help ease the isolation and loneliness of rural living.

Other users may include social service agencies which lack their own communications facilities. For example, in Guyana the Inter-American Development Bank (IDB) uses the MEDEX system extensively for ordering building materials for construction of new health facilities in the rural areas where the medex are located. Messages from the IDB are relayed to MEDEX headquarters in Georgetown, where they are telephoned to the IDB staff.

Still other users might include businesses, private individuals, church groups, and various government agencies. The number of applications of the system depends directly on the operating protocols established by the health system. Given the generally acknowledged repressed demand for communications services in rural areas, planners should expect that there will be great pressure for using the health system for all sorts of general communications functions.

Disaster Relief

Two-way communications systems are also earning an impressive reputation as a fast and inexpensive as well as highly mobile way of dealing with emergencies such as epidemics and natural disasters. In Guatemala during the 1976 earthquake which devastated many areas,

PLENTY, a voluntary organization, was able to set up a two-way radio system which was one of the first emergency communications systems to operate during the disaster. The system linked relief teams, ambulances, and local health workers with hospitals, clinics, and fire-emergency units in the hard-hit lake country of Guatemala. In most villages, the radios were the only link with the outside world.

The ATS-1 satellite network in the Pacific has been used to coordinate emergency medical services and logistics during outbreaks of cholera and dengue fever in the South Pacific islands.

In the southern U.S., hospitals communicate with oil rigs in the Gulf of Mexico and with a helicopter "ambulance" during medical emergencies via the ATS-3 satellite.

Problems and Sources of Failure of Two-Way Radio Projects

Technical Problems

While it is often difficult to establish the reasons for failure of two-way radio systems, there are several prominent areas where system failure can easily occur. It should be noted that health organizations are rarely prepared to undertake the long-run maintenance of radio systems. Unless the radio system is very large and the health organization well funded, it is difficult to justify employing a technical staff for the system's maintenance. As a result, maintenance, planning, and equipment replacement must often be left to others--consultants, equipment manufacturers' representatives, or telecommunications ministries.

Regardless of the source of operational maintenance, its existence must be established and regular. As with any technology, long-term neglect can lead to catastrophic failure of the equipment with the requirement of a more costly replacement rather than less expensive continuing maintenance. Available evidence indicates that major causes of system breakdown or failure are likely to be:

● Poor system design.

● Insufficient training of users or inadequate operational procedures.

● Lack of spare parts or foreign exchange to purchase spare parts.

● Power supply problems.

Faulty system design occurs most often when attempts are made to minimize initial capital costs. Such savings may be achieved by purchasing older or used equipment or by purchasing simpler equipment, such as Citizens' Band radios, in the belief that they will provide the necessary service.

An example of the first can be seen in a project in South America where the project director initially requested that army surplus radio equipment be used. Such equipment, using vacuum tube technology, would have meant considerable initial capital savings, but would also have meant continuing operational problems compounded by the difficulty of obtaining spare parts and by the costs of the larger power consumption of vacuum tube transceivers compared to transistorized equipment. Fortunately, the project engineer in this case convinced the funding agency that purchase of modern, solid state equipment was a considerably more cost-effective investment.

In another project, in Nicaragua, Citizens' Band radios were purchased for a two-way health project. While the initial testing of this equipment led to positive results, the equipment is currently unusable over the distances required because of sun-spot activity and will probably always provide questionable service. In this case, the more costly investment in HF-SSB (high frequency-single side band) systems probably would have proven more cost-effective over the life of the project.

Operational procedures are also a source of radio problems. Such problems range from operators being careless with handling the microphones or placing objects over the heat exhaust of the transceiver to lax security and placement of transceivers in poor locations such as areas exposed to moisture or heat, dust, or salinity. For example, in Guyana, the main station of the MEDEX radio system was directly exposed to humid salt air blowing in off the ocean. Without remedy, this radio would be corroded beyond repair in less than two years.

Lack of training of operators can result in early damage to equipment if radios are allowed to burn out or corrode, batteries are not charged properly, battery cables to battery terminals are reversed, or gasoline is not properly mixed with oil for generators. Even if original operators are trained, the radios may soon be used by other untrained operators.

Maintenance protocols are a major consideration in the success of two-way radio systems. While solid-state radios should operate for long periods of time without the need for maintenance by a technician, some periodic preventive maintenance is advisable.

Lack of routine maintenance may severely shorten the life of the system. For example, in Lesotho, it was found that many of the antenna installations needed repair as antennas had broken, radios were not grounded, and poles were leaning dangerously. In the Pacific it was found that moisture in long grass around the guy wires for the poles rusted the couplings holding the guy wires. The guys broke, the poles fell over, and the antennas came down. Cutting grass can extend the life of a radio system!

In the absence of such maintenance procedures, systems can totally disintegrate. For example, the Ethiopian Malaria Eradication Program's radio system apparently failed for lack of spare parts. In Liberia, radios were repaired sporadically, and because of lack of proper management procedures, they were diverted from government repair shops to other projects.

One reasonable approach to this problem, which has been used by AID in Guyana, is to contract with the local telecommunications authorities for installation and maintenance. In many projects, there is real reluctance to use the telecommunications authorities often because of their operational record, a traditional lack of cooperation among ministries, and the cost. However, the telecommunications authority usually is best suited to maintain the radios because of its own supply of skilled technicians, and often the location of its repair shops near health headquarters.

However, the Inter-American Development Bank will not fund dedicated two-way radio projects when there is a possibility of having the telecommunications ministry install the facilities. Among the reasons cited for this are maintenance requirements--IDB is concerned that equipment cannot be effectively maintained on a haphazard basis as occurs when a user ministry attempts to provide communications service.

Lack of spare parts can seriously threaten any technology including two-way radio. Some systems are procured with funds just sufficient to acquire the original equipment. When spares are required, they may be difficult to obtain, and/or the government may have insufficient foreign exchange to buy them. This factor contributed to the demise of a network in Nigeria.

Power supplies are a persistent technical problem. In locations with existing power source (for example a town power supply or generator for the hospital), surges in voltage can damage the radio. Voltage regulators can be included to prevent damage. However, where voltage is much below specified output, it may not be possible to use local power to run the radio or recharge its batteries.

Self-contained power units for two-way radios usually consist of small diesel generators used to recharge automobile storage batteries. The generators must be cleaned and maintained properly. Another problem is the cost and transportation of diesel fuel. The logistics of transporting diesel oil to remote locations can often be horrendous. For example, in mountainous areas of Lesotho, in interior Guyana, and in the Arctic, fuel must be flown in.

In contrast, solar panels should eliminate the need for generators and fuel, although at present their capital cost is higher. Field tests do not indicate any major problems with solar panels, but none have been in operation long enough for definitive evaluation.

Organizational Problems and Source of Failure

A second set of causes of system failure or abandonment relates to the perceived utility of the system. To be valued and used by the rural health worker, the radio must provide communication not only to peers but to referral points such as regional hospitals and to administrative centers for ordering of drugs and supplies. The evaluators of the Liberian Lofa County Rural Health project communication system noted that the failure to operationalize the communication network between the levels of care as outlined in the project proposal had several serious repercussions.

226

Referrals, particularly for emergency cases, cannot be effected and lives are consequently often needlessly lost, in light of the absence of a rapid, dependable means to notify the appropriate higher level of care of the need for technical advice, supplies, and/or an ambulance to transport a critically ill patient to the hospital.

Physican Assistants (PAs) remain professionally isolated both from colleagues and, importantly, the physician on whom they are dependent for back-up. Such isolation has an obvious detrimental effect on PA motivation and dedication to service.

Lack of support and continuity for field facilities encourages patients to bypass lower levels of care in favor of overcrowded county hospitals where all illnesses can be treated and referral is not necessary.

In Lesotho, health planners and nurse practitioners pointed out that it was extremely important for each clinic to be able to communicate with its regional hospital. (The variety of privately owned systems each with separate frequencies now makes this difficult.) In Guyana, the medex who can now communicate with headquarters in Georgetown, and with each other, pointed out that communication with regional hospitals for referrals and supervision would make the system more valuable for them.

Conclusions

This paper has presented an overview of the applications of two-way communication in rural health care delivery and the various communications technologies that can be used. The main focus has been on two-way radio systems as an appropriate technology for dedicated health communication networks in developing countries.

The main functions of two-way communications in the projects reviewed are medical consultation and referrals, administration, coordination of transportation, and continuing education. Where radio networks are shared with other users in rural communities, a variety of applications for other development sectors may be used.

The cost of two-way communication systems is

modest, varying from several hundred to a few thousand dollars, depending on the distance to be covered, the terrain, and the availability of a power source. Operating costs also depend on a number of factors including proper training of users, regular maintenance, the cost of fuel, and the organizational structure responsible for maintenance.

Several lessons can be drawn from the projects outlined in the paper.

● The communication service must meet the perceived needs of the user to be accepted.

● The equipment must be simple to use and reliable.

● System budgets must include adequate allocations for operations, maintenance, and spare parts.

● Training for operators in proper use of the radios and in preventive maintenance may extend the life of the system.

● Regular operating schedules and procedures may increase the effectiveness of the system.

● Educational applications require a much greater investment in personnel time to prepare programs and courses than do consultative and administrative applications.

Questions

1. Contrast communications networks owned by user organizations and contractors.

2. Which important sociocultural factors should be considered in selecting a communications system?

3. What factor actually influences several inappropriate choices of equipment?

4. What are the implications of this statement from context: "Cutting grass can extend the life of a radio system!"?

Case No. 16

Technology Choice in a Fisheries Project

by
Paul Bundick,
World Bank, Washington, D.C.

This case highlights overall technological choice
and the selection of equipment in the light of existing
constraints. Also, assessment of the appropriateness
of the choices made, especially with reference to national development goals.

TECHNOLOGY CHOICE IN A FISHERIES PROJECT

Introduction

This case examines the choice of technology in an
investment project that was undertaken to improve the
efficiency and production of the fishing industry in a
low-income island developing country. It should bring
the reader greater insight into the relationship between technology choice and the domain of feasibility.
The case should also suggest that there are instances
where the domain may be very narrow, constricted by
severe constraints and lack of opportunities so that
the project designers are directed to the selection of
one technology by a clear lack of alternatives. In
other instances the domain is more extensive, allowing
for the consideration of a wider range of technological
options and more "room" for the project planners to decide on the basis of cost-benefit comparisons and to
bring into play their value preferences in choosing
the "best" technology among a range of economically
viable options. The case provides opportunity to

This case was prepared by Paul Bundick in September 1983 using World Bank documents and additional information provided by Mario Kamenetzky (PAS) and Gert
Van Santen (ASP), who critically reviewed preliminary
drafts.

assess the appropriateness of the overall technological choice (export of fresh fish) and the selection of equipment (motorized vessels) in the light of the existing constraints and opportunities.

Background

The project described in this case was carried out in a nation consisting of hundreds of small coral islands remotely situated outside the mainstream of world commerce and communications. The islands range in size from tiny islets barely above water at high tide to more substantial islands reaching several kilometers in length. Because they are based on coral, the islands are flat. The highest elevation is less than 3 meters above the sea. The vegetation is tropical, typically coconut palms towering above dense scrub brush. Crystal-clear lagoons marked by multi-colored reefs and an abundance of tropical fish and marine life of every shape and color provide a spellbinding spectacle of natural beauty.

The islands are grouped together in clusters called atolls. These atolls form a long narrow chain covering an ocean area of more than 100,000 square kilometers. Most of the islands are situated close to a reef which surrounds each atoll. Inside the reef, waters are relatively calm and protected. Nearly all of the atolls have lagoons that afford anchorages for vessels of medium draft. The enclosure reefs to each atoll contain openings which provide passageways for small sailing boats traveling between atolls. Only a few reef openings, however, are deep and wide enough to accommodate even small modern cargo ships.

The people who inhabit the atolls form a closely knit society, unified and disciplined by common bonds of religion, language, and culture. A strong sense of kinship and family bonds contribute to social cohesion. The inhabitants of each island form an interrelated group where everyone knows everyone else, where social services, law and order, and investment decisions are the responsibility of the community directed by the island's chief. The atolls and some of the larger islands have traditionally been self-contained social and economic units, dependent on the sea around them, as fishing is the main economic activity.

For many years, the nation's economy has been based on the harvesting of fish for consumption and export. Tuna are abundant in the island waters with

the main species being skipjack and yellow-fin. Skipjack and juvenile yellow-fin are surface tuna and feed
year-round in the waters outside the atoll reefs. Deep-
swimming tuna, primarily mature yellow-fin, migrate
through the ocean channels during two to six months of
the year. Numerous species of bottom fish and reef
marine life abound in the atoll lagoons. These include
among others: Grouper, red snapper, several species
of white fish (red seabream and kingfish), many varieties of sharks, and small pelagic baitfish. Lobsters
and limited quantities of shrimp are found in the reef-
protected waters.

The fishing industry has traditionally operated
on a relatively small scale with harvesting limited to
local waters. Approximately three-fourths of the fish
landings comprise skipjack and yellow-fin tuna. Each
atoll has its distinct season and catch characteristics. Only a few atolls sustain regular catches
throughout the year.

The prevailing fishing techniques in the country
are the result of hundreds of years of fishing experience. Inside the atolls, the water depths rarely exceed 100 meters. Outside the reef, the ocean floor
drops off to a depth of 2,000 meters or more. Tuna
rarely enter the lagoons, and fishing for these species
takes place outside the reef in open water up to a
maximum distance of 25 kilometers. The lack of navigational aids and "fishing customs" discourage the
fishermen from attempting to fish beyond the sight of
land. In addition, wind variability and dangerous cur-
rents in the channels deter traditional sailing boats
from venturing too far into the open sea in search of
tuna. Changing wind conditions increase the risk that
the crew will have to row their fishing vessel back to
port. Rowing more than several kilometers against the
current with a boatload of tuna is a difficult and
time-consuming task.

The fishing boats, called pole-and-line vessels,
are made of coconut palm wood and waterproofed with
matting and shark oil. They vary in length from 8 to
14 meters and employ a crew of 8 to 13 men.

A fishing trip begins in the early morning when
the vessel sails to the baitfishing area inside the
coral reef. Baitfish are attracted with fish paste
and caught in fine-mesh nets.

When sufficient bait is collected, the vessel

231

clears the reef and heads for deep water in search of tuna. When the tuna are sighted, the pole-and-line vessel sails alongside the school and live bait is thrown overboard sending the tuna into a feeding frenzy. Lines are lowered with barbless hooks into the churning waters. Tuna are then caught and pulled on board in a carefully controlled manner so that the hooks disengage just before the fish hits the deck.

When the bait is used up, the fishermen return to shore with their catch. The tuna are stored in the baitwell or on the deck, periodically splashed with water to keep them as cool as possible in the hot tropical sun. The catch is usually sold on the same day it is caught to avoid spoilage. The fish is consumed fresh or processed into salted, smoked, and dried fish for storage and/or export. Over 90 percent of the country's tuna is caught using this pole-and-line method.

The citizens of the country have one of the highest per capita fish consumption rates in the world, fluctuating from 55 to over 90 kilograms per person per year. The people show a marked preference for tuna over other kinds of fish. Some types of fish and crustaceans, considered delicacies in Europe and the United States, i.e., lobster, snapper, shrimp, etc., are rarely eaten by the local population. About half of the food consumed is purchased from other countries, this being rice, flour and sugar. Domestic production, largely fish, cereal crops, tubers, coconuts, and vegetables and fruits, supplies the other half.

Agriculture has traditionally played a minor role in the island's economy. The soils are highly alkaline and lack many essential nutrients, particularly nitrogen. The coral sand-based soils are also very poor in moisture retention, a fact which limits the types of crops which can be grown.

Coconuts are also an important natural resource and form a major part of the local diet as well as the raw material for much of the handicraft industries and boat-building trades. On some islands, poor nut production can be traced to the tough underbrush which competes with the palms for scarce nutrients and water resources. Nut production could be increased in certain areas if the brush were cleared among the trees. This underbrush, however, provides the people with their only major source of fuelwood--about 56 percent of the country's total effective energy utilization

and 95 percent of the cooking fuel. The annual consumption of fuelwood nationwide is 50,000 MT (metric
tons) per year. Generally, only the tourist resorts
and about 2 percent of the households in the capital
city use kerosene for food preparation. The vast
majority remain dependent upon firewood as their primary energy source.

Fish and marine products account for almost all
merchandise exports in the traditional economy. In
the 1960s, by far the most important export item was
"Native Fish"--a specially processed form of skipjack
tuna.

Native Fish is an indigenous product made by use
of a simple technology. Skipjack tuna, ranging from 3
to 10 pounds, are cut into pieces and boiled in salt
brine. The tuna are then smoked on racks made of bamboo erected over wood fires. After smoke-curing, the
fish are further dried in the sun. Following the drying process, the cured tuna is bagged in gunny sacks
for export. One kilogram of Native Fish requires the
processing of 5 kilograms of fresh tuna. A nutritious
by-product is also derived from the process--a liquid
fish stock high in nutrients.

Native Fish is produced in small-scale home industries exclusively by women. When cured, the product is stored in individual households until it can
be transported and sold to the STO (State Trading Organization) in the capital city.

In 1967 the export of Native Fish represented
over 90 percent of the total export trade. In the
past, the product has only been exported to a neighboring developing country where it is used as a condi-
ment with traditional foods.

For the last several decades, the country's main
exports and imports have been handled by the State
Trading Organization, a semi-autonomous public organization connected with the Ministry of Finance. The
STO provides the major source of revenue for the national budget. The government allows the STO to use a
special administrative official accounting rate in foreign exchange transactions enabling it to buy cheap
from local producers and reap high profits when selling the same produce in overseas markets. The commercial (free) exchange rate of the national currency (NC)
fluctuates around U.S. $1.00 = NC 8.80. The official
accounting rate is fixed by the government at U.S.

$1.00 = NC 3.93. The STO, for instance, buys Native
Fish from the local fishermen using the administrative
accounting rate and sells it at the international market value of the national currency. Thus the fishermen only receive from 40 to 60 percent (depending on
currency fluctuations) of the international market
price for their product. This is equivalent to taxing
the fishermen, the primary occupational group, nearly
half of their gross earnings in order to subsidize food
imports.

The STO uses the revenues generated in the fish
exports to partially subsidize the cost of essential
food imports.

Transformation of the
Fishing Sector in the 1970s

The fishing sector has traditionally dominated
the national economy providing nearly all export earnings and a livelihood for the vast majority of the
island population outside the capital city. Annual
fish landings from 1967 to 1978 fluctuated greatly,
varying between 22,600 metric tons (MT) in 1968 to
35,400 MT in 1974. The growth of fish landings was
slow, averaging only 1.8 percent per annum during that
decade, well below the rate of population growth.

This poor performance conceals a major structural
transformation in the fishing sector which began in
the early 1970s. In 1971 the sale of Native Fish, the
principal export commodity, reached a record 5,400
metric tons (27,700 fresh tuna equivalent), an increase of 40 percent over the 1967 level. In 1971,
this accounted for about 95 percent of the country's
exports.

In 1972, owing to a foreign exchange crisis, the
only buyer (foreign) of Native Fish cut back sharply
on imports of this commodity. That year, sales of
Native Fish dropped to 3,800 metric tons and even
greater reductions appeared likely. The government
responded quickly to this decline in exports by invit-
ing foreign companies operating in adjacent waters to
start purchasing fresh fish for freezing and export to
markets in other parts of the world. An immediate response came from three companies. Two were wholly
foreign-owned, while one was a joint venture with the
government, the foreign holdings accounting for 60 percent of the shares. After price negotiations, the companies decided to set up mechanized collection systems

to buy fresh tuna. In that first year, they purchased
2,000 metric tons of fresh tuna. Although the start
was slow, it established a trend that transformed the
fishing sector in the years that followed.

The decline of the Native Fish market continued
throughout the decade, falling to only 240 metric tons
in 1978, just 4 percent of the 1971 level. In these
same years, the sale of fresh tuna continued to increase especially after 1975. In 1978 the sale of
fresh tuna to foreign companies accounted for two-
thirds of fish exports and 44 percent of the total fish
landings. These changes had significant implications
for the economy as a whole.

The sale of fresh tuna to foreign companies required the creation of a new marketing infrastructure
to collect and store the fish prior to transporting it
to far-away markets. The companies provided collector
vessels to pick up the catch from different atolls and
mother ships where the tuna could be frozen and kept
in cold storage while awaiting transfer to other ships
for export. By 1978 this infrastructure consisted of
six mother ships with a combined cold storage capacity
of 2,700 cubic meters and a daily freezing capacity of
160 tons. One company operated a 30-ton cold store on
one atoll.

In 1978 there were nine collector vessels operating in the atolls. Unable to cover all the islands,
these nine ships concentrated on the most promising
areas for fish collection. In order to sell fish to
the foreign companies, a local fisherman had to deliver
his catch to one of the collector vessels or directly
to the mother ship where the tuna could be graded,
weighed, and receipted in the presence of an STO official.

Since many of the atolls have good catches for
only two or three months out of the year, some fishermen were forced to travel great distances in order to
sell their tuna to the company ships that operated only
near the atolls with the highest tuna yields. Local
pole-and-line sailing vessels had difficulty in knowing
the exact location of the collector vessels or the
mother ships and had extreme difficulty in following
and reaching the motorized company ships.

In the era before 1972, the fishing activities of
nearly 2,000 sailing vessels were spread out virtually
over the entire country. The change in tuna marketing

introduced strong pressures for the powering of traditional vessels. In 1974 the government decided to motorize four fishing boats on a trial basis. In a short time these mechanized traditional vessels, retrofitted with 22 horsepower diesel engines, proved their versatility and efficiency in comparison to the sailing boats. Motorized vessels were able to follow and easily reach the collector vessels and the mother ships and thus sell their tuna to the export market. Installation of engines on traditional crafts proceeded rapidly so that, by 1978, about 700 of the country's 2,150 pole-and-line vessels had been retrofitted with diesel engines. In that year the mechanized fleet contributed over half of the total tuna landings and nearly all of the fish exports to foreign companies. Many of the traditional sailing vessels, which before the decline in exports had operated for the Native Fish market, were forced to greatly reduce their operations pending motorization because of the lack of marketing opportunities.

This mechanization of sailing vessels was largely confined to boats operating out of the capital city and atolls immediately to the north where fresh fish collection was mainly taking place. In the southern atolls, the retrofitting of boats lagged behind the national average; by 1978 only 18 percent of the southern pole-and-line vessels had diesel engines as compared to the 30 percent countrywide average and the 55 percent for the capital city.

Fuel for the motorized vessels soon became a problem with recurrent fuel shortages and distribution inefficiencies disrupting fishing operations. To insure the collection of tuna, the companies began to distribute diesel fuel to local fishing vessels at a subsidized rate. The government's normal channel of fuel distribution was through the STO, which priced diesel fuel at NC 1.65 per liter as compared to the foreign company subsidized rate of NC .60. The subsidized fuel was allocated to the fishing boats according to the quantity of fish delivered to the collector vessels or mother ships. Because the companies bore all the cost of the subsidy, they tended to restrict the sale of fuel to what they believed was the absolute minimum necessary to conduct fishing operations. Many vessels, running short of fuel, were therefore forced to travel to the capital city in order to purchase additional fuel from the STO. Fuel imports increased tenfold between 1975 and 1978 reflecting the mechanization of the fishing fleet and the rising demand from the tourist industry for energy and transport. The fastest growing import item in the 1970s was petroleum, reaching 9.2 percent of the country's total imports in 1978.

The international tuna market in the 1970s was highly competitive with numerous brokers and canning companies competing for the fresh fish supplies. Demand for skipjack, the major tuna used in the canning industry, continued to rise throughout most of the decade. Between 1972 and 1979, the average price of fresh skipjack paid by the foreign companies to the STO rose from U.S. $167 to U.S. $255 per metric ton, or at 6.1 percent per annum, while the world market skipjack prices rose from U.S. $446 to U.S. $816 per metric ton, or at 9 percent per annum. The difference between the world market price and the prices paid by the foreign companies to the STO are partially explained by the higher percentage of smaller tuna in the country's catch and the high cost of collection, freezing, storage, and transport borne by the companies. The slower rate of increase of prices paid to the STO, however, does reflect the relatively weak bargaining position of the country's government vis-a-vis the companies. The price received by the STO has been estimated at 50 percent below the market value.

The rising costs (primarily of fuel) and the unchanging selling price in the early 1970s caused many fishermen to begin producing dry-salted fish (some tuna, but primarily different species of reef fish) and dry shark fins for export. Both of these new export products were comparatively high value non-perishables and neither of them was subject to an STO monopoly. In 1977 the government began to require that private exporters of dry salted fish convert half of their foreign exchange earnings at the administrative accounting rate, thus introducing a tax on these exports, albeit at a lower rate. Dry salted fish and dry shark fins increased faster (as a percentage of the total exports) than selling of fresh fish to foreign companies. The salted fish is produced locally in home industries, like the Native Fish, but it is not smoked. The export trade is primarily in the hands of local private companies which sell the product to a nearby developing country. The dry-salted fish market potential is not well known but is expected to be limited and substantially below the market for fresh tuna. The rise in tourism has also stimulated a small domestic market for reef fish, shrimp, and lobster.

In 1977 a small tuna canning factory (15-ton capacity), a gift from a bilateral assistance program, was
installed 100 kilometers north of the capital, on an
island with some water resources in a prime yield location. It is operated by the same joint venture company
(60 percent foreign, 40 percent local) involved in the
purchase of fresh fish for export. The factory is dependent on importing cans and, unlike larger factories
in the industrialized nations, does not use the significant amount of dark meat and fish scraps left over
after the light meat is canned. Larger factories have
developed alternative product lines (e.g., pet food)
for the high amount of waste material in the canning
process. Fifteen tons of tuna will produce only about
two to three tons of canned product.

Social and Economic Changes
Occurring in the 1970s

The country's population estimated at 103,800 in
1967 rose at a rate of 3 percent a year to reach
147,000 by 1978. Local religious law permits contraceptive measures if these do not result in sterility.
Custom and religious tradition, however, favor large
families by positive injunction. This population is
well distributed throughout the island chain; only 19
islands have more than 1,000 inhabitants. One striking
exception to this distribution pattern is the capital
city where over one-fifth of the total population of
the country is crowded into a small island of only 2
square kilometers.

The capital is the nation's only urban settlement.
In the last quarter of a century it has come to domi-
nate the political and economic structure of the country. A few families make up a small national elite
owning many of the fishing boats and leasing them out
to fishermen on the atolls. (This trend has been increasing in recent years especially with the motorization of boats.) This elite benefits from a dispropor-
tionate share of the government's expenditures, consum-
ing much of the subsidized imports. Generally, the
residents in the capital enjoy a substantially higher
standard of living than the inhabitants of the outer
atolls. The outer atolls rely upon the capital as
their main trading post and contact point with the
rest of the world.

In recent years, migration to the capital has become a serious problem. An estimated 40 percent of the
city's population in 1977 was classified as "non-

resident"--people lured to the city by the concentration of health and education facilities and possible job opportunities. In the same year, unemployment reached 15.3 percent of the active labor force.

In many ways the urban elites are cut off from the traditional social structure. Not only are the elites more receptive to foreign ideas and techniques but also they tend to tolerate greater deprivation in their city environment than would be accepted in the self-regulating atoll communities.

Other rapid changes occurring in the 1970s included an impressive rate of increase (12.5 percent a year) in Gross Domestic Product (GDP). This fast growth was attributed to the expansion of the money economy in general and the emergence of tourism and shipping as major economic activities in addition to fishing.

In 1978 tourism accounted for nearly 12 percent of the GDP and for the first time surpassed the fishing sector as the number one earner of foreign exchange. Shipping also contributed to the diversification of export earnings during the 1970s. National Shipping Ltd. (NSL) is a government-owned shipping company which provides the main regular link between the islands and the rest of the world, and in the same year accounted for 22 percent of government revenues and 14 percent of gross foreign exchange earnings.

In 1978 fishing accounted for over 20 percent of the GDP and 45 percent of the employment, excluding the sizable catch in the non-monetized economy. Fishermen were by far the most important occupational group, numbering nearly 20,000 in the 1977 census. In the same year, 5,850 women listed their primary occupation as fish processing. This was about 25 percent of the female labor force and 9 percent of the country's total labor force.

The Fisheries Project/
The Choice of Technology

In the late 1970s, the government undertook preparation of the country's first development plan. The plan did not articulate specific development objectives but did formulate four medium-term goals:

1. A rate of economic growth sufficient to ensure an adequate increase in real per capita income and

consumption;

2. A more equitable distribution of the benefits of development as between the capital city and the outer atolls;

3. A determined effort to slow down migration to the capital through the creation of alternative poles of development; and

4. A program to reduce the rapid population growth.

   Recognizing the importance of fishing for the economy but also the waning production, the government targeted the sector as a primary area of investment. A survey of the sector was carried out, leading the government to formulate a plan with the following two objectives: (a) To increase the production of fresh tuna for export through foreign companies, and (b) to improve the efficiency of the fishing operations.

   A project team then undertook a preliminary engineering study of the improvements to be made. An analysis produced the following information:

(a) The motorized traditional vessels were found capable of catching about 30 tons of tuna per year as compared to an average of 10 tons for a non-powered sailing boat. The motorized boats were able to reach the fish schools much faster and waste less time in traveling to and from the fishing waters. The number of potential fishing days was also increased by 15 percent because the fishermen were no longer at the mercy of wind variations.

(b) Crews in pole-and-line vessels work for a share of the catch. This share varies according to the atoll and according to the kind of boat being used. Typically the crew's share of a sailing vessel is 60 percent of the catch divided equally among the crew members (after the cost of the fishing gear is subtracted). Extra shares are given to the skipper and the master fisherman. The owner of the boat receives the rest.

(c) A typical share of a crew on a motorized vessel is 40 percent of the gross catch less fuel costs and engine repairs. The owner receives the remainder after the traditional extra shares for the skipper and master fisherman.

(d) The estimated annual gross sales and operating

costs of the two types of vessels at 1978 prices were
calculated as follows:

	Mechanized	Non-mechanized
Gross sales	28.5	9.4
Operating costs	21.1	6.9
Fuel	8.7	-
Wages	7.9	5.6
Maintenance	2.8	0.8
Other	1.7	0.5
Debt service on engines	3.7	-
Net sales proceeds	3.7	2.5
Depreciation	2.7	0.7

(e) For a typical mechanized vessel, the wages in the
form of catch sharing amounted in 1978 to NC 7,900 for
a crew of 10 or an average of NC 790 per year per person. The average wage for a crewman on a sailing craft
in 1978 was NC 560.

(f) The net income accruing to the average proprietor
of a motorized vessel was about NC 3,700 per year,
whereas the owner of a sailing boat earned NC 2,500.
This figure did not allow for depreciation, which is
substantially higher for mechanized owners (NC 2,700
per annum) than for traditional sailing vessels (NC 700
per annum).

(g) Under optimal circumstances, mechanization enabled
a 40 percent increase in the income of fishermen and an
approximately 50 percent increase in the net income of
boat owners.

The study also suggested that the advantage of
motorization could be enhanced by enlarging the size of
the vessel.

The study also revealed that collector vessels and
mother ships could operate only during the daylight
hours because of treacherous reefs and the lack of
navigation aids. Several collector vessels had already
struck reefs in their efforts to pick up fish. Local
fishing vessels were also found to be hampered by lack
of navigational aids especially with the increase in
the number of motorized vessels operating in unfamiliar
waters.

Based on this pre-engineering study, the project
team selected the following components, which they

considered would be required to achieve the project objectives:

1. Credit to boat owners for the motorization of 500 sailing vessels (425 traditional and 75 new improved models as a pilot effort);

2. The establishment of five maintenance and repair stations;

3. The installation of buoys and reef markers;

4. The development of five prototype motorized vessels to be designed by interested local boat-makers for future manufacturing; and

5. Technical assistance and a follow-up study.

The project was to be implemented over a period of 2½ years at a total cost estimated at U.S. $3.89 million.

A feasibility analysis examined cash-flow projections, financial cost-benefit ratios and estimated economic rates of return for traditional sailing crafts, motorized traditional vessels and new motorized vessels. This led to selection of the motorized option as the best way to increase productivity.

The plan's financial benefits consisted primarily of the value of the incremental catch resulting from vessels motorized under the project and improved efficiency of the vessels because of the navigational aids and reef markers which, it was assumed, would allow for early morning departures and late night returns of the fishing boats. With project implementation, the annual incremental landings were expected to yield about 11,900 tons of skipjack valued at U.S. $3.25 million. The estimated financial rate of return from the motorization of an existing traditional sailing vessel was 24 percent. The estimated rate of return for a new motorized vessel was approximately 29 percent. In establishing the financial rates, the official exchange rate of NC 3.93 per dollar was used.

The economic rates of return for the motorized vessels, both newly built and retrofitted traditional crafts, were calculated to be much higher than the financial rates of return. With various adjustments to the financial projections, the economic rate of return for investments in the motorization of existing sailing vessels was over 100 percent. For investments in the

242

motorization of new vessels, it was even higher.

Questions

1. In 1972, at the beginning of the Native Fish market decline, what alternative overall technologies would you have considered as possible options to the export of fresh tuna to foreign companies? Discuss how the constraints and opportunities (at that time) would bear on each one of these alternatives.

2. In what ways do you think the decision to motorize traditional sailing vessels added or detracted from the national development goals listed in the text?

3. In 1979 when the project was being designed, the decision was already made to base the development of the fishing sector on the selling of fresh fish to foreign fleets. Hence the project cycle skips the stage of the selection of an overall technology and enters at the pre-engineering stage where equipment and organization are selected. Could you think of alternative equipment or additional installations which could have been considered besides the three basic boat types and the navigational aids in order to increase fish exports?

4. In addition to the alternative equipment in Question 3, can you think of alternative organization and management systems that would contribute to achieving project objectives and the national development goals referred to in Question 2?

5. Discuss the effects of the government's pricing policy on the fishing sector, its implications for the success of the project objectives and national development goals.

6. What research projects would you consider most essential to the development of the fishing sector?

Case No. 17

Papa China (Taro) Project, Colombia

by
Barbara Myers,
Appropriate Technology International,
Washington, D.C.

> This case describes how one technological innova-
> tion--extracting starch from taro root for human con-
> sumption--led to its diffusion to another agricultural
> sector--hog production. Also, how the particular tech-
> nology to achieve this result was selected and the pro-
> duct evaluated.

Background

The project which was originally proposed to and
approved by Appropriate Technology International (ATI)
was to establish a community-based commercial plant to
extract starch from taro, a tropical root crop known
locally as "papa china." The plant was to be located
on land owned by four communities near Puerto Merizalde
on the Maya River, about six hours by boat from Buenaventura, on Colombia's Pacific Coast. The market for
the starch was to be the textile industries in the
Cauca Valley. The project concept postulated that existing technologies for extracting starch from yucca
could be successfully transferred and adapted to taro.
This assumption was based on information provided by
the Foundation for Industrial Development, a non-profit
agency of the Cali Chamber of Commerce. Under Foundation auspices, a Colombian agricultural consulting firm
had conducted successful laboratory tests of starch extraction from taro.

When additional technical information became
available, the advisability of basing the entire project on this one technology was brought into question.
The Tropical Products Institute (TPI) in England, in a
reply to an inquiry by the director of the Instituto
Matia Mulumba (IMM) in Colombia, stated that it knew
of no successful field testing of starch extraction

Abridged and reprinted with the permission of
A.T. International.

methods for taro. The TPI advised IMM's director, Father Miguel Angel Mejia, as to the likelihood of encountering problems because of the smaller size of taro starch granules and getting lower yields because much of the starch remains in suspension. The other problem cited in TPI's letter was the fact that taro contains a considerable amount of mucilaginous material which tends to clog the strainer used in the simple precipitation method of starch extraction.

ATI had hired a consultant to assess the project from a technical standpoint and to assist IMM in preparing a project plan. The consultant visited IMM shortly after TPI's letter arrived and together with Mejia it was quickly decided to shift the project focus to explore several alternative applications for taro in addition to starch extraction.

The revised version of the project included two phases. The first phase consisted in exploring the technical and economic feasibility of producing several alternative derivative products from taro root. The second phase was the actual commercial operation, when ownership of the physical assets of the project would pass to a mixed corporation which would produce whichever product was identified during the first phase as having the greatest potential. The principal objective of the project was to expand the community's ability to generate income. Thus, derivative products of taro needed to be identified--products which were marketable at relatively high unit values. Four products were found which met this requirement:

Taro starch. The development of an economic system for extracting starch from taro root remained the highest priority since refined starch is the product with the greatest potential value.

There are two methods for separating starch from taro. The simpler is by decantation, using gravity to draw the starch through a filter from a taro and water solution. Tests were run in the laboratory facilities of the University of Valle in Cali using different strengths of taro solution and containers of varying dimensions. These tests yielded a much lower separation rate for taro as compared to yucca starch, but the method still appeared to be feasible. When subsequent tests on a commercial scale were conducted at a yucca processing plant in Dugua, the major part of the starch particles remained in suspension and this line of investigation was abandoned.

246

There was some research data available which indicated that the addition of small quantities of aluminum
sulfate enhanced the precipitation process, but this
was not pursued beyond the laboratory stage since the
method was not considered to be appropriate to local
conditions. Problems associated with dosage and handling of the chemicals by local residents who were untrained in such techniques were considered to be too
risky.

The other method of starch extraction uses centri-
fugation to separate the particles from solution.
Laboratory tests using a conventional centrifuge resulted in rapid separation, but this method also had to
be abandoned when no equipment of an intermediate size
or technical level could be identified. The centri-
fuges used in commercial yucca processing are all designed for large-scale operations, requiring high motor
force and advanced technology. Therefore, starch extraction was determined unfeasible for commercialization given the project conditions.

Taro chips. There is a known and substantial market in Europe for chips made from yucca, which are included in dairy products and hog rations as a source of
carbohydrates. It was postulated that if taro chips
could be produced, it would be possible to market them
either domestically or internationally. To make chips,
the taro is first washed and then fed into a chipping
machine which slices the root into thin pieces. These
pieces are then dried and packaged.

For this sub-project a Malaysian-style pedal-
powered chipping machine used for yucca was acquired
with project funds and modified with the assistance of
ASTIN, a local program funded by the German government
to aid small industry. Technical problems which arose
during the trials included the incomplete processing
of the taro root, the non-uniform size of the chips,
and the clogging of the blades. In addition, the
machines became disaligned after a few hours of operation. These problems were at least partially overcome
by further modifications in the design, but the drying
operation was never successful because of the problems
cited below. Therefore, taro chips were determined to
be impractical for commercialization given the project
conditions.

Taro flour. There is local demand for taro flour
as a partial substitute for wheat flour. Processing
consists of cleaning, dehydrating, and pulverizing the

247

taro root. A wood-fueled drying chamber was constructed for this sub-project, as well as for the chips sub-project. Because of the extremely high humidity of the area, moisture was reabsorbed from the air to such an extent that the taro became soggy. In some cases mold even grew on the taro before the dehydration process was complete. Therefore, taro flour was determined impractical for commercialization given the project conditions.

Hog production. Taro root, properly prepared and balanced, can serve as the principal source of carbohydrates in hog rations. The domestic market for pork in Colombia is virtually unlimited and the potential conversion (value-added) factors appeared to be favorable. Therefore, hog production was investigated further as the most promising in terms of its commercialization potential.

Before it can be consumed by either animals or humans, taro must first be cooked in order to break down the oxalic acid which causes inflammation of the mouth lining. To prepare the hog rations, the taro is chopped and then boiled and mixed with a protein supplement--either soymeal or fishmeal--since taro alone does not contain enough protein to promote adequate growth. A commercial vitamin and mineral supplement are also added to insure a balanced diet.

Project funds were used to purchase land in Puerto Merizalde where yield tests for different densities were conducted and sources for taro were sought out. Yields ranged from 46 to 80 tons per hectare without any fertilizer or other soil additives being applied, thus confirming the optimistic projections on which the project was based. This land also served as a demonstration plot, and local women (responsible for agriculture in the local culture) were taught how to plant taro in rows, with optimum spacing between the plants.

Because the major expense involved in feeding taro to hogs is the soy cake that is added as a protein source, the project is now exploring the possibility of producing fishmeal locally at low cost from trash fish or refuse from commercial fishing operations. This fishmeal would replace soy as the principal source of protein in hog feed, and thus reduce factor costs even further.

Technology Dissemination
and Commercialization

The immediate impact of the project will be within five small communities located near Puerto Merizalde. Approximately 500 people will be directly involved in raising hogs, with the specific arrangements for sharing work and responsibilities to be determined by each community. There are considerable opportunities to disseminate this technology beyond the Puerto Merizalde area, both to other communities where IMM is working as well as through other organizations active in the Pacific Coast region. IMM prefers, however, to allow the commercial phase of the project to mature before actively promoting it elsewhere. It considers that the work to date has been a success in that it has shown that it is technically and economically feasible to raise hogs on taro-based rations. Many of the details of the commercial operation still have to be decided and, more importantly, proven in practice, before the model is used elsewhere. Similarly, there are human and cultural factors which may have an important impact on the ultimate success or failure of this activity. The coastal population is accustomed to a subsistence lifestyle and previous experiences with introducing organized economic activities have not always met with success.

Project Evaluation

A number of factors can be identified as contributing to the project's success.

The Nature of the Sponsoring Organization

External technical resources. IMM's ability to draw on external technical resources was crucial to the making of correct decisions at several points in the project's evolution. The technical advice provided by the Tropical Products Institute prevented the project from being pursued along lines which almost certainly would have resulted in failure.

The consultant from ATI hired to assist IMM in developing the project plan also provided technical input which was a key to the project's success. He had previously worked with taro and other tropical root crops in the Caroline Islands in the Pacific, and it was at his suggestion that hog raising was included as one of the four lines of investigation in the project. The consultant later joined ATI's staff and assumed responsibility for monitoring the project. In this capacity he was able to provide on-going technical advice to the project coordinator in such areas as determining the appropriate mix of ingredients for the hog rations, deciding when to abandon other lines of investigation, and formulating a record-keeping and reporting system to insure that wider learning from the project experiences would occur.

Once hog raising was determined to be economically viable, ATI made additional technical resources available to IMM through the services of a Colombian consulting firm which helped identify alternative commercialization strategies and an accounting system for the community-based enterprise. IMM considers this input to have been invaluable since no one on its staff had the training to plan the business aspects of the project.

In-house technical capacity. The taro project was able to avoid the pitfalls of many of IMM's previous agrotechnical projects partly because a technical coordinator was hired to be in charge of project implementation. As an agronomist he had the technical background necessary to conduct the various experiments, but he was also familiar with the region since he had grown up in Buenaventura. He readily established a good rapport with the local residents and kept them regularly appraised of the progress being made in the project. IMM is of the opinion that without hiring a technical coordinator, the project could never have been attempted since no one on the staff had either the knowledge or the time to provide the kind of ongoing supervision required. Thus, hiring the coordinator insured that there was an appropriate match between the technical requirements of the project and the organization's capabilities.

Ability to retain organizational autonomy. Another factor related to the nature of the sponsoring organization was the working relationship established between the project coordinator and IMM's director. Mejia allowed the coordinator total autonomy in decisions related to the project, even when the two disagreed. This proved to be important when IMM underwent its series of leadership changes because the project was able to continue relatively undisturbed by Mejia's departure and the organizational crisis which followed.

End-user involvement. From the beginning there

was considerable risk inherent in the project. The technical nature of the research phase also meant that there were limited ways in which the community could be directly involved. Nonetheless, a consistent effort was made to insure that local residents, and in particular the leaders of the communities which would be part of the commercialization phase, were aware of the objective and evolution of the project. Locating the testing facilities in Puerto Merizalde itself despite its inaccessibility enabled local residents to drop by out of curiosity and to become familiar with the project and its progress. The site was prepared and the test station was built with the participation of local carpenters and workmen. The construction design has served as a model for several families which have imitated it and used local materials in building their own homes. The end result was an effective approach to the problem of project risk and the involvement of the end-user community. The community felt itself to be a part of the project (i.e., the project was "theirs"), but they did not directly assume any of the risk.

The State of the Technology

It is also clear that a number of factors related to the nature of the technology itself contributed both to the success of the hog raising sub-project and the failure of the others.

Tolerance and simplicity. The techniques involved in the hog raising technology are extremely simple and require only a minimum amount of instruction in how to prepare and measure the hog rations. The tolerance of the technology to occasional errors in mixing the proportions or in the amount of feed given is also favorable. The consequences of an occasional mistake in mixing the feed are negligible. The most likely problem to arise is underfeeding, which would simply result in slower weight gain but would not mean failure of the whole effort. These features are expected to be essential to the project's ultimate success because of the low educational level of the local population.

Use of local resources. One of the main characteristics of the technology is that it relies almost entirely on local resources. Taro itself is one of the few crops which is resistant to insects and which is able to withstand the heavy rainfall, periodic flooding, and extremely acidic soil conditions in the region. Its appropriateness to local conditions also stems from the semi-migratory lifestyle of much of the

251

population, since the harvest date is not determinant. It takes eight months for a taro crop to mature, but if it is left in the ground for several additional weeks or even months it can still be harvested.

The pig pens, with the exception of the cement floors, are built entirely out of local materials such as bamboo and palm trunks. Only by relying on locally produced inputs, especially given the remoteness of the region, can costs be kept low enough to make the effort profitable. The projected substitution of locally produced fishmeal for soy in the hog rations will be a further improvement in the technique because it will increase reliance on locally available resources and thereby reduce input costs even more.

In contrast, the sub-projects which were abandoned were found to be incompatible with local conditions. High humidity made any process involving drying (chips, flour) impractical. The problems encountered in starch extraction could only have been overcome by using technologies which either did not exist on an intermediate scale (centrifuge) or relied on inputs with low tolerance to error in handling (chemical reagents).

Questions

1. Explain this statement from context: "There are human and cultural factors which may have an important impact on the ultimate success or failure of this activity" (to raise hogs on taro-based rations).

2. Explain the process of technology selection with reference to taro-based rations. How were the alternative techniques for extracting starch from taro root eliminated?

3. How would you assess the vantage point of the narrator with respect to project evaluation in this case?

Case No. 18

Communicating with Pictures, Nepal

by
National Development Service, Nepal,
and
United Nations Children's Fund (UNICEF), Nepal

This case indicates why psychological and sociocultural differences must be considered in the selection of pictorial messages in technology transfer since
every picture tells a story, but not the same story.

Most Nepalese villagers cannot read. So they
naturally cannot understand booklets or other written
material about village development.

- Is it possible to communicate ideas and information
  to villagers by using pictures only?

- What kinds of pictures are most meaningful for villagers?

- Do different colors have special meanings for villagers?

In early 1976 the National Development Service and
UNICEF conducted a study designed to get answers to
these questions. Teams of data-collectors went to nine
different parts of the country: the mountains, hills
and plains in the Far West, West/Center, and East. They
conducted interviews with over 400 adult villagers from
the following groups: Thakuri, Brahmin, Chhetri, Mus-
lim, Bhote, Magar, Gurung, Maithili, Limbu, Rai, and
Tharu, showing them a wide variety of pictures and
colors and noting their responses. None of the villagers interviewed had ever been to school.

Some of the pictures used in the study.

Findings of the Study

Note:
1. In the following explanation, the term "villagers" is used to refer to the sort of villagers interviewed in the study, i.e., adult villagers who live away from the main towns and who have never been to school (which means most of the villagers in the country). Villagers who have been to school, or who have had other opportunities to look at many pictures, would probably find the pictures easier to understand than the villagers who were interviewed in the study.

2. The pictures used in the study measured approximately 17 cm. x 13 cm.--i.e., they were all bigger than they appear in this study.

Question: Is is possible to communicate ideas and information to villagers by using pictures only?

Answer: Probably not.

In the course of the study, over 20 pictures intended to convey ideas (rather than just to represent objects) were shown to villagers.

Many (but not all) of the villagers could recognize the objects shown in the pictures. But the ideas behind the pictures were almost never conveyed to the villagers.

For example, the picture on the next page was intended to convey the idea that people who drink polluted water are likely to get diarrhea. It was shown to 89 villagers, and only one of them understood the message behind the picture.

This series of pictures was intended to show how to
make rehydration mixture for the treatment of diarrhea:

The interviewers told the villagers the meaning of the writing in the second picture (salt, baking soda, sugar), but the series was not comprehensible for any of the 89 villagers who saw it.

Why Pictures Fail to
Convey Ideas and Information

1. Villagers who are not used to looking at pictures may find it difficult to see what objects are shown in the picture.

   "Reading" pictures is easier than reading words, but people do not automatically know how to do it. People have to learn to "read" pictures.

   In the final stage of the study, just one picture was taken from the series shown above.

   It was shown to 410 villagers. Only 69 of them realized it was a picture of hands putting something into a pot. Ninety-nine others could see the hands but could not suggest what they might be doing. And the rest of the villagers (242 people) did not see the hands at all. Eighty-two of them thought it was a picture of flowers or a plant.

Sixty-seven villagers were shown this picture, but only 33 of them saw that it was meant to be of a person digging. Many others thought it was a man with a gun, and others suggested it could be someone flying a kite.

This picture is supposed to represent a pregnant woman, but about a quarter of the 410 villagers who saw it did not recognize it as a woman. Eleven percent thought it was a man, and others thought other things such as a cow, a bird or a rabbit.

Villagers in the least-developed part of the country (the Far West) had the most difficulty in understanding the pictures, probably because they see fewer pictures than villagers in other parts of the country. For example, over 400 villagers were shown 10 pictures of fairly common things (such as a horse, a woman, a water pot). On average, the villagers in the West/ Central and Eastern areas recognized about seven of these 10 pictures, but the villagers in the Far West recognized an average of about four of them.

2. Villagers do not expect to receive ideas from pictures.

Four hundred and ten villagers were shown this picture. Ninety-nine of them realized it was meant to be a picture of someone drinking boiled (or heated) water. But even when the interviewers asked what the picture might be trying to teach, only 10 of these villagers suggested that "drink boiled (or heated) water" might be the message. Many of the villagers were surprised at the suggestion that a picture could teach anything.

3. Villagers tend to "read" pictures very literally.

That is, even if they recognize the objects or people represented in the picture, they may not attempt to see any link between the objects, or any meaning behind the picture.

This picture of a funeral pyre was meant to represent the idea of death. Some of the villagers said they could see a person lying on firewood, but they were not prepared to state that the person was dead, or to suggest any other explanation of what they could see.

4. Villagers do not necessarily look at a series of pictures from left to right, or assume that there is any connection between the pictures in a series.

For the final stage of the study, an attempt was made to simplify the "diarrhea cycle" picture, as follows.

Less than half of the 410 villagers looked at these pictures in order from left to right (37 percent of them looked at the middle picture first). Hardly any of the villagers appeared to think the pictures were related to each other.

5. Pictures which try to convey ideas or instruction often use symbols which are not understood by villagers.

The arrow is a symbol very commonly used to indicate direction. (For example, it has been used in three of the drawings reproduced in this study so far.) But very few villagers think of an arrow as indicating direction.

These symbols are often used to mean "right" and "wrong", or "good" and "bad". But the villagers interviewed for this study, not having been to school, had not learned to understand these symbols.

This sign is often used to represent danger. But only 4 out of the 410 villagers knew it was supposed to indicate danger. (However, over half of the villagers recognized it as a skull, or said it was a monster or a ghost, or remarked that it gave them an unpleasant feeling.)

Question: Are pictures any use then?

Answer: Yes.

People are interested and attracted by pictures, even though they may need help to interpret them. If a picture's message is explained to villagers, they will probably remember the message when they see the picture again.

For example, this is a picture used by the British-Nepal Medical Trust to illustrate the idea that TB can pass from the lungs of a sick person to the lungs of a healthy person.

During the study, this picture was taken to six villages and shown to over 100 people. In 5 villages, none of the villagers who saw the picture could understand it. But in the sixth village, many villagers could explain exactly what the picture meant. They could understand it because 5 months before, some health workers had visited their village and talked about TB, and had shown them this picture.

262

Question: What kind of pictures should be used?

Answer: Realistic pictures, with a minimum of background detail, are the easiest to understand.

The NDS/UNICEF study tested ten pictures in six different styles:

Two styles of photograph:

1. Photo with background
2. Block-out (photo without background)

Four styles of drawing:

3. Shaded drawing
4. Line drawing
5. Silhouette
6. Stylized drawing

1.

3.

2.

4.

5.

6.

The block-outs were usually more effective than the photos. For example, 67 villagers saw this photo, and 79 percent of them recognized it.

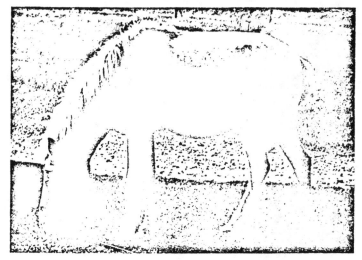

Sixty-six other people from the same village saw this block-out, and 92 percent of them recognized it.

If results of all the ten pictures are taken together:

- Block-outs were recognized by 67 percent of the
  villagers.

- Photographs were recognized by 59 percent of the
  villagers.

The shaded drawing was the most effective style of
drawing (in fact, it was a little more effective than
the block-out).

If results of all ten pictures are taken together:

- Shaded drawings were recognized by 72 percent of the
  villagers.

- Line drawings were recognized by 62 percent.

- Silhouettes were recognized by 61 percent.

- Stylized drawings were recognized by 49 percent of
  the villagers.

Some examples:

Sixty-six villagers saw this shaded drawing of
a mountain and 56 percent of them recognized it.

267

Sixty-nine other villagers saw this line drawing and only 28 percent of them recognized it.

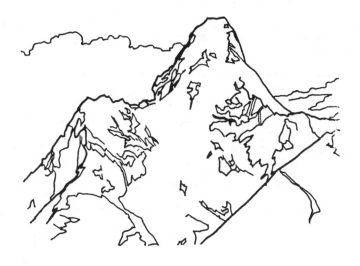

Sixty-eight other villagers saw this stylized drawing and only 16 percent of them recognized it.

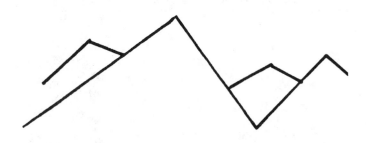

But some stylized drawings can be quite effective.

For example, 410 villagers saw this drawing, and 65 percent of them realized it was meant to represent people.

Sixty-eight villagers saw this picture, and 75 percent of them understood it was meant to be a water pot.

Sixty-six percent of the villagers interviewed recognized this as a drawing of a fish.

Four hundred and ten villagers saw this drawing, and 59 percent recognized it as the sun. Fourteen percent said it was the moon.

Even very simple drawings are recognized best if they are of things very familiar to village people.

In areas where houses have sloping roofs, 85 percent of the villagers recognized this as a building. But in areas where houses have flat roofs, only 26 percent knew it was meant to be a building

Question: What kind of pictures should be avoided?

Answer: Pictures showing a lot of different objects are not well understood.

Eighty-nine villagers were asked what they could see in this picture. Only 53 of them recognized the houses, even though they are quite realistically drawn. Far fewer of the villagers mentioned any of the other objects in the picture.

Silhouettes are often misinterpreted. People are inclined to think a picture like this is a black person,
or a monster, or a devil, or a ghost.

What About Colors?

The villagers interviewed in the study liked
bright colors, and thought of them as colors for
happy occasions. The most-liked colors were red,
bright pink and purple. Orange and yellow were well
liked, and people thought of them as the colors for
gods. Blue and green were also liked, particularly
in the Far West. The villagers interviewed disliked
black very much, and thought of it as the color for
devils. They also disliked brown and grey.

What Should You Do?

What should you do if you want to use pictures
to help you communicate ideas to villagers as you
carry out your village development work?

1. It is possible for you to make simple drawings
    yourself that will be quite effective--certainly
    more effective than just talking or writing. It
    is not necessary for you to use photographs, which
    are difficult and expensive to make.

2. The most effective drawings are shaded drawings
    with little or no background, like the one shown
    on the following page. However, these are the

most difficult to draw.

3. If you cannot make shaded drawings, or get someone
    else to make them for you, then try to make line
    drawings, like the one on the following page.

4. If these are still too difficult to make, then make stylized drawings, like the one shown on the following page.

5. If you decide to use colors, choose colors which
    are appropriate to the ideas you are trying to
    communicate:

        bright colors are for happiness;

        orange and yellow are for gods;

        black, brown and grey produce negative
        reactions.

    If you color your drawings, make things the colors
    they are in real life, or villagers will probably
    be confused.

6. Make your drawings of such things as houses, water
    pots, people, etc., as much as possible like the
    houses, water pots and people that villagers see
    every day, or they may not recognize them very
    easily.

7. Do not put too many objects in one drawing. Each
    drawing should contain only one or two objects, if

possible. It is better to have many drawings with one or two objects in them than to try to put many things in one drawing.

8. Do not expect villagers to learn a lot from the drawings alone. Use drawings to capture the villagers' attention, to reinforce what you say, and to give them an image to remember, but always give a clear and full oral explanation of your subject in addition to showing the drawings.

9. If you want to place drawings around the village to continually convey ideas to villagers (e.g., signs indicating "danger" or "poison", or reminders about family planning, or conservation and reforestation, or the need to boil drinking water), then remember it will be necessary to teach many villagers what the drawings mean. It is quite possible to do this teaching, e.g., by getting all children at the school to make copies of the drawings and take them home and teach their families what they mean. Unless many villagers are taught the meaning of such drawings, the drawings are not likely to be effective and may even have a negative effect if they are misinterpreted.

10. If you use a series of pictures to represent an idea or a set of instructions, be sure that the villagers see the relationship between the various pictures in the series (e.g., that the pictures all represent the same person at different stages of some action). Make sure also that the villagers know which order to "read" the pictures in.

11. Remember that villagers will be likely to interpret your drawings very literally, e.g., if you draw something larger than it is in real life (such as drawing a fly six inches high), people may assume you really mean it to be an impossibly enormous fly, or they may think it is a strange kind of bird.

12. If posters are sent to the village by one of HMG's Departments, e.g., the Agriculture Department, or the Family Planning/Maternal and Child Health Project, you can help by making sure the villagers understand the meaning of these posters.

Questions

1. In your estimate, which is the more difficult problem to solve: The technical transmission of messages or the psychological, social, and cultural problems involved in their understanding?

2. Suggest remotely analogous cases of simple graphic messages that are not understood at all or misread even in developed societies (you can begin with traffic signs, instructions on do-it-yourself gadgets, labels on cans and bottles).

3. Compare this case study with the PIACT mini case in Chapter 3 above on the use of contraceptive pills. Both relate to the transfer of technology through pictorial messages. Do you perceive any differences?

Chapter 5
Technology Transfer and Public Policy

These cases focus on technology transfer decisions by public authorities, whether these are the direct recipients of the transferred hardware or software or merely the regulators and/or promoters of transfers to the private sector. Ponder over the possible motivations of the public administrators--including sociocultural or psychological ones--in each of their decisions.

Case No. 19

A Publicly-Owned Steel Mill

by
Paul Bundick and Robert Maybury,
World Bank, Washington, D.C.

This case highlights the policy of local capacity expansion but also the neglect of human values during project implementation. It contrasts the success of a technical transfer on one hand with the inappropriate treatment of management and social problems on the other, adversely impacting on productivity, environment, and thus economic-social cost-benefit.

Introduction

This case study summarizes the design and early implementation of a large public investment project in a middle-income developing nation. One of the country's development priorities was to promote self-sufficiency in steel production, a goal that would accelerate the growth of the capital goods industries and also meet the projected domestic demand for steel by eliminating the need for imports. An additional consideration stressed by national planners was the need to create new poles of development in the economically backward areas of the country. In response to these priorities, the government created a public enterprise charged with formulating a specific plan for a modern steel plant in one of the nation's underdeveloped regions.

_____

This case was prepared by Paul Bundick and Robert Maybury in October 1982 on the basis of materials and comments provided by John Jaffe (IND), Mario Kamenetzky (PAS), and J. M. H. Tixhon (PAS), who critically reviewed preliminary drafts.

The project was initiated in the early 1970s under the management of a local public enterprise but with the technical and operational assistance from an overseas steel corporation. State-of-the-art technology and many components were purchased from a number of advanced industrialized countries and installed by a team of engineers from the local public enterprise working with a group of foreign experts from the overseas corporation. Corporation experts were placed at various management and specialist levels throughout the project to facilitate the transfer of skills to the local enterprise.

Economic considerations, engineering design, and the selection and installation of hardware are key factors in the success of any large capital investment project. Success requires, however, that equal attention be given to developing local technological capacity to operate, maintain, and manage technology as well as to the human problems associated with the mobilization of large numbers of workers during the implementation of the project. In describing some of the successes and problems encountered during project implementation, the case study raises some of these issues, especially the organization and efficient transfer of engineering skills from a foreign to a local team. The case also throws light on the importance of ecological and human factors in project design. If neglected, environmental and social problems may hinder an otherwise well-planned implementation.

The Setting

The investment project described in this case was undertaken in the early 1970s in a middle-income developing nation. During the post-World War II period, the country achieved rapid and sustained economic growth with relatively low rates of inflation. In 1970 the GNP per capita climbed to U.S. $600 after increasing for more than 20 years at a rate exceeding 3 percent per annum.

Much of the country's impressive growth can be attributed to the dynamic expansion of the industrial sector. Since the mid-1950s, industrial growth averaged more than 8 percent annually. During the 1960s, the share of manufacturing in the labor force rose from 16 to 23 percent. The most important product groups in terms of annual growth rates were textiles, engineering goods, chemicals, petrochemicals, and metals. The growth of output in the last four product groups reflected government's attempt to diversify the sector as well as the country's developing engineering base.

Despite these high rates of growth, the country's economic expansion was accompanied by serious income disparities and regional imbalances. National industries were primarily established in urban areas where there was a greater demand for their products and where infrastructure was already developed. The lure of employment led many in the rural areas to migrate to the cities. The high rate of population growth (over 3.4 percent per annum) and the relative scarcity of arable land in many parts of the country also contributed to the rapid increase in the urban population. The urban population since 1960 rose more than 5 percent annually and in 1970 accounted for 60 percent of the total population. In turn, the growing abundance of cheap labor was attractive for establishing new industrial enterprises in and near the expanding urban zones.* This pattern resulted in the heavy concentration of economic activity in a relatively few dynamic growth centers.

The unequal pattern of development was further exaggerated by geographical and climatic factors. High mountain ranges and desolate arid zones have historically isolated certain regions from other parts of the country. Regions also are unevenly endowed with natural resources. In the past, highways and railroads were primarily built to serve the resource-rich zones and urban centers in the more accessible fertile valleys and highland plateaus. In the 1960s the government sought to overcome these regional discrepancies by promoting more geographically balanced growth. Priorities were given to new industrial projects in underdeveloped areas.

During the three decades of rapid growth, the government's development planning placed heavy reliance on import substitution in the capital goods

_____

*Unemployment was recorded at around 9 percent in 1970 while an additional 30 percent of the economically active population were employed in activities of such low productivity as to be considered underemployed. Moreover, because of the increasing population and its young age structure, the labor force will continue to grow very rapidly.

industries* and the mobilization of both public and private sector resources to achieve their economic goals. Over the years they sought to create a regulatory and institutional environment conducive to allowing private and public enterprises to compete. This included favorable tax structures and considerable import protection. The government also played a major role in creating essential infrastructure, introducing incentives and promoting key industries such as power, petroleum, and steel. The construction sector, stimulated by government public works and housing programs, grew at an annual average rate of over 9 percent in the 1960s. The expansion of both industry in general and the construction industry in particular was the primary factor in the nation's growing demand for steel.

Steel provides one of the basic materials needed for the growth of the industrial sector. During the 15-year period from 1954 to 1968, the nation's apparent consumption** of steel products soared at an average annual rate of 10.3 percent, reaching 2.6 million tons in 1968.*** The apparent consumption of non-flat products (bars, rods, and structurals), much used in the construction industry, increased fourfold in the same period accounting for 47 percent of the total apparent consumption in 1968.

In 1969 growing concern about the future supply of steel and the prospects of increased imports toward the end of the 1970s prompted the government to appoint a survey team to study the steel sector and assess the need for future investment projects.

Results of the Steel Sector Study

The study of the steel industry revealed that the nation's total nominal production capacity was expected to reach about 5.2 million tons in 1972 without additional plant construction. The production capacity

---

*Not until the early 1970s were important steps taken to develop manufacturing exports on a large scale. In 1970 industrial exports were 5 percent of the manufactured output.

**Apparent consumption equals domestic production plus imports less exports.

***This is equivalent to about 3.5 million tons of raw steel.

would be distributed among 10 public and private companies operating both integrated and semi-integrated
facilities.

The survey team determined the actual raw steel
production to be about 20 percent under the nominal
capacity. Technical limitations,* imbalances among
production units, shortages of scrap iron for the semi-
integrated producers, and poor location of some plants
in relation to raw materials and markets all contributed to the discrepancies between nominal and actual
production. The members of the research team appraised
the effective raw steel capacity of the nation's plants
to be around 4.2 million tons per year.

The installed capacity for rolling finished products was found to exceed the nation's steel-making
capacity. The survey team estimated that by 1972 the
effective rolling capacity of the steel industry would
be 2.4 million tons of flat products, 1.8 million tons
of light non-flat products, 0.6 million tons of heavy
sections, and 0.2 million tons of seamless pipe for a
total of 5.0 million tons. To fully utilize this
finishing capacity, the industry would require 6.7
million tons of raw steel input, or 2.5 million tons
above the estimated raw steel production for 1972.

In their analysis of steel demand, the survey team
forecast a steady increase in the nation's consumption
of steel products. Taking into account the government's plan to continue its housing and public works
program started in the 1960s, they projected that the
total finished product demand would rise 10.4 percent
per year from an estimated 1972 base of slightly less
than 3.1 million tons (4.2 million tons of raw steel).
The demand was expected to reach 6.7 million tons of
finished products (9.2 million tons of raw steel) by
1980.

The survey team concluded from the available data
that the nation's nominal capacity was insufficient to
meet growing demand in the 1970s. Even computing the
nominal capacity at 5.7 million tons (in accordance
with existing plant expansion plans) and increasing
the utilization effectiveness of the nominal capacity
from 80 percent to 85 percent, the team projected defi-
cits as early as 1974 with increasing steel imports

---

*Only a small portion of the pig iron being pro-
duced was by state-of-the-art methods.

required in the years ahead.

Excluding structurals from their calculations, the
survey team estimated that the demand for light non-
flat products (bars, rods, and sections) would grow at
an annual rate of 9 percent from a 1972 base of 1.2
million tons, reaching 2.8 million tons by 1980. The
projected supply for light non-flat products for the
same period was considered to be sufficient in 1972
but was expected to fall steadily behind throughout
the remainder of the decade. The team calculated that
by 1980, without any increase in the production capacity, the country would have to import about 1.0 million tons of non-flat products.

The availability of raw materials needed for
steel industry expansion was assessed by the study to
be sufficient, with the exception of high-quality coal
for coking and occasional shortages of imported scrap
iron. While coal was found in the northern provinces
of the country, it was high in ash content (16 percent) and of too low a quality for cost-effective iron-
making.

At the time of the study, the country was nearly
self-sufficient in iron ore, importing only 1 percent
of its total need. Manganese ore and other alloying
elements required in steel production could be purchased easily within the country. Abundant domestic
supplies of natural gas and petroleum were also available in several areas. In 1969, however, the full
potential of these reserves had not yet been explored.

The existing steel industry in the country was
protected by tariff barriers. The study noted that
the import duties on steel products averaged about 20
percent. Steel products sold in the domestic market
by national producers were also subject to price ceil-
ings. Since the 1950s, domestic market prices were
usually lower than ceiling prices and no distortions
resulted from the price controls. In anticipation of
increases in production costs during the 1970s, the
study estimated that prices would reach 2 to 3 percent
above ceiling levels by 1976.

In their sector analysis, the survey team outlined
three alternative options to fill the steel demand/
supply gap projected for the late 1970s:

Option 1 - Importation of semi-finished steel products
            (billets and slabs) for domestic rolling

into finished products;

Option 2 - Production of steel in electric furnaces, based on imported sponge iron and scrap, followed by the rolling of products; and

Option 3 - Fully integrated steel production using domestic or imported iron ore.

In their analysis of the first option, the team concluded that it would not be practical for the country to meet future steel demand by importing steel billets and slabs. Supplies on such a large scale were judged too expensive and difficult to obtain because of the unpredictability of world supplies. At the time of the study, long-term contracts for semi-finished steel could not be secured on the world market.

The study also concluded that the second option was unattractive because of the volatility of the world scrap market and the high opportunity costs of using large quantities of the country's electricity. At the time of the study, the steel industry imported an average of 600,000 tons of scrap per annum. As mentioned, shortages in scrap periodically reduced the efficiency of the semi-integrated producers. Although the study concluded that there was a continuing role for scrap-based increases in production capacity, particularly for specialty steel, the unreliability of world supplies made this option unattractive for large-scale operations.

The survey team supported the third option as the most viable route to meet projected demand increases. They recommended that a new site be selected instead of expanding the capacity of existing producers. The existing producers were all located inland and none was adjacent to large unexploited reserves of iron ore required by a large-capacity steel plant. If a new plant were to be viewed simply as a means of adding a million tons to the nation's steel capacity, the team stated in their report that marginal expansions of existing producers would probably be a cheaper option. But when the expansion plans were taken into account along with the government's development objectives, the short-term capital cost advantage of expanding existing plants was superseded by the benefits of a new steel mill. They recommended that the government invest in a new steel plant preferably constructed near the coast so that the new mill may be able to take advantage of raw material imports and steel exports through a port.

287

The Project

In 1969 the government, on the basis of the sector study, decided to conduct the pre-investment work related to the design and construction of an integrated steel plant. It set the following two objectives for the future project:

(1)   To meet projected increases in domestic non-flat steel demand with a modern steel works that could later be expanded to produce a full range of steel products at competitive world market prices; and

(2)   To direct public investment towards underdeveloped areas and away from the concentrated centers of wealth and economic activity.

The guidelines set by the government stated that the project was to be carried out in several stages stretching over 15 to 20 years. The first stage would provide a plant installation and the infrastructure needed to produce and market one million tons of light non-flat steel products (wire rods and light bars and shapes). A second stage would begin producing steel plate in four to five years' time. Later expansions into flat steel production were planned to yield an estimated 5 million tons per annum. To carry out project design and implementation, the government founded a public company which we will call the State Enterprise and which would be in charge of the construction and operation of the proposed plant.*

A location was selected for the proposed project in a relatively underdeveloped coastal zone with available raw materials. Iron ore was found in primary deposits within 20 kilometers of the proposed plant site. Reserves, according to estimates, could be mined cost-effectively for over 30 years. Barium oxide, copper, silica sand, limestone, and bentonite were also available near the proposed site. The surrounding region also had extensive timber reserves and deposits of

---

*The government owned 51 percent of the shares which were legally non-transferable, assuring continued federal ownership and control. The remaining shares were divided between a national development bank, a government-owned steel company, a federal trust fund, and private interests.

dolomite clay used for refractories. Coal reserves,
however, were not available in the region even in small
low-grade concentrations found in other parts of the
country. Petroleum and natural gas were also lacking
in the region, with the country's main reserves located
hundreds of kilometers from the proposed site.

The plant was to be constructed near a small town
of 10,000 people. The mill would be located on the
coast, near a potential deep-water port facility and
adjacent to a large navigable river. The selection of
the site was also influenced by a powerful political
figure who was committed to bring progress through industrialization to his underprivileged native state.

In 1969 the region was primarily rural with more
than 60 percent of the population engaged in agriculture. Production levels were generally low. The region's industry produced primarily footstuffs, textiles, and various chemicals. In 1969 it employed
8 percent of the region's population and generated a
per capita output of U.S. $21 as compared to an average
U.S. $240 per capita for the country's industrial sec-
tor as a whole. About 30 percent of the population
depended on construction, commerce, and services to
earn their livelihood. At the time of the proposed
project there was insufficient surplus labor available
in the vicinity to meet the demand for a large construction project. In the immediate area there was
only a population of about 15,000 people.

The major reason for the state of underdevelopment in this region was its isolation from other parts
of the country. At the time, the nearest railway link
was 200 kilometers and the closest port was 250 kilometers away. Two large dams had been built several
years before on the major river system in the region.
One dam was located only about 15 kilometers upstream
from the proposed site. The region's two hydroelectric plants accounted for 13 percent of the country's
installed capacity. In addition, the region provided
opportunities for maritime industries, and the beautiful coastal beaches offered a good potential for the
development of tourism.

At the time of the proposed project, the goal of
producing one million tons of steel products in an integrated plant leading eventually to an expanded production of 5 million product tons per year could be
reached by one of two basic processes: The conventional blast furnace/basic oxygen furnace (BF/BOF) and

the less conventional direct reduction/electric furnace (DR/EF) process. In 1969 the BF/BOF process was firmly established as an effective iron and steel-making technology. The DR/EF process was less widely known and had not been utilized on a large scale, even though the manufacture of ordinary grades of carbon steels in electric arc furnaces had been fairly common in some countries since 1942. Direct reduction with natural gas and the DR/EF combination had been commercially available since 1954. At the time of the project, at least four different processes of direct reduction were in different stages of development, one in use in the country's own private sector accounting for about 20 percent of the country's steel capacity. Even so, the DR/EF process in 1969 did not yet enjoy complete worldwide acceptance because of its limited commercial application.

These two processes yield equal quality steel in the manufacture of light non-flat products but each process has its own distinctive characteristics and advantages. The conventional BF/BOF process requires large quantities of high-grade coal (to make coke used in the production of pig iron). This process also entails use of large units (yearly capacities of several million tons), thus allowing economy of scale advantages. The coke ovens associated with this process, however, cause substantial pollution unless environmental controls are installed and maintained.

The DR/EF process requires high-grade iron ore and large amounts of natural gas. Coke is not necessary for the DR/EF operations but the steel-making stage of this process requires abundant electric power. The DR/EF process generally emits somewhat less pollution than the BF/BOF method and is more compact and easily controlled since it is usually limited to smaller-scale works with outputs ranging from about 250,000 tons to one million tons annually. No technical constraints, however, prevent larger plants from being constructed.

The State Enterprise, after an analysis of the two overall technology options (in relation to the proposed site's constraints and opportunities), decided on the BF/BOF process. The pre-feasibility study led to the decision to proceed with the preliminary engineering and the selection of specific components for a new and modern integrated steel plant that would exploit the iron ore deposits in the proposed site area.

In the early stages of the project design, the

290

State Enterprise conducted international competitive
bidding and selected a consortium of overseas steel
manufacturers (Overseas Steel Corporation, or OSC) to
assist in the design and implementation of the plant
construction and subsequent operation. Under an agreement signed with OSC, the State Enterprise retained
managerial control and responsibility for project execution but assigned to the OSC an obligation to provide
assistance in all technical, financial, marketing, and
organizational aspects of the project. The plan specified that OSC staff were to be integrated with local
counterparts and placed in various managerial and
specialty positions throughout the State Enterprise because of the comparative inexperience of many local
engineers.

This agreement was in many ways a model agreement.
It provided for a real transfer of skills in project
design, implementation, and operation. The foreign
experts were to work as a team with local engineers
hired by the State Enterprise and to provide direct
assistance in building local engineering capabilities.
The agreement also included provisions for training
all levels of the labor force and explicitly stated
that all the major engineering work would be performed
in the project country.

In the pre-engineering stage, the joint State En-
terprise/OSC team selected the project components and
formulated a specific plan for the purchase of the
state-of-the-art BF/BOF steel technology. The plant
engineering design included the mining and benefication
of iron ore, a pelletizing plant, conventional coke
ovens, a lime plant, a blast furnace, a basic oxygen
steel-making plant, the continuous casting of billets,
and two rolling mills. The rolling mills were to be
designed to allow sufficient flexibility for a product
mix to meet the full range of likely market requirements. The plan called for the plant layout to be
constructed in a way that made it easy to expand and
diversify production facilities.

The plan called for steps to minimize air, water,
and noise pollution usually associated with mining and
steel plant operations. The dust and gas emissions of
the major facilities (coke ovens, pelletizing plant,
lime plant, blast furnace, steelmaking plant, etc.)
were to be controlled by increasing the height of the
smokestacks beyond the necessary limits to better dis-
perse combustion emissions. Extreme noise generated
by the blast furnace (often attaining averages over

100 decibels) was to be limited to set levels. Specifications negotiated with the equipment suppliers set the maximum noise level at 85 decibels at a distance of one meter from the operating furnace. Water pollution control was given priority because of the recreational and commercial potential of nearby beaches and off-shore waters. The polluting effect of discharge water used in cooling and cleaning operations was to be reduced through recirculation of the processed water. The water would be returned to the sea after treatment to reduce solid particles and chemical residues.*

The steel plant, when its first stage reached full production, would import about 765,000 tons of coal per annum. Therefore, the project plan included the construction of a port at the mouth of the navigable river on which the steel plant was to be built. The port capacity was to accommodate bulk vessels and deep-water berths for the importation of coal. Ten percent of the steel production was to be exported after the port's completion in 1977. To provide transport for finished products to the domestic market, a railway connection (about 200 km.) was to be built to link the plant with the national railway system. A road would be constructed by the time of production start-up and would serve to transport steel products to the major urban centers until the railroad's completion in 1978.

The project included a large component for the recruitment and training of both managerial and operational employees. Allowing for attrition and turnover in the labor force, the project planners estimated that a total of 5,000 management, maintenance, and production personnel would require training from 1972 to 1978. To accommodate this need, a training center was to be built at the plant site with approximately 80 instructors to train the labor force for the construction and operation of the steel mill. In addition, a management team of approximately 200 engineers and foremen were to receive training overseas and also on-the-job through their expatriate counterparts.

The training program was geared to a successful start-up and a smooth transfer of skills and responsibilities to the management of the State Enterprise. Every process department was initially assigned an OSC

---

*Direct operating costs of pollution control devices were estimated at U.S. $2 per ton of product.

manager and several technical personnel. The function of these expatriates was to create an operating team and train their State Enterprise counterparts on-the-job. The OSC staff would then be phased out once the State Enterprise's personnel could assume full managerial and supervisory functions.

At full production in 1976, the plant's total operational labor force was projected to reach 4,000 employees including head office and sales personnel. This was expected to climb to 8,000 by 1996 in accordance with the planned expansion. Indirect employment created by the demand for supplies and in the purchasing of consumer goods and services by the mill's employees was projected to total an additional 8,000 in 1976 reaching nearly 24,000 by 1996. Assuming the national average of 3.7 dependents per employed worker, the total population of the "steel company town" would reach 100,000 by the year 2000.

By 1973 the planners estimated that more than 1,700 workers would be involved in the construction of the plant facilities. This labor force would increase to 7,000 by 1974 and reach its peak of 9,300 in 1975 before leveling off to the 4,000 regular operational personnel in 1976. In view of the shortage of available labor, the initial work force had to be recruited and relocated to the plant construction site. This influx of thousands of workers meant that housing and other facilities had to be constructed. Although building of housing and commercial facilities was planned for the long term, temporary housing would have to be constructed at the initial stages of project implementation. Because of the housing limitations and the temporary nature of the facilities, the workers would be asked to share quarters and were not to be permitted to bring their families until the completion of plant construction in 1976. The planners also felt that the expatriate engineers should be discouraged from bringing their families because of housing shortages and so that all of their attention could be focused on the speedy and efficient completion of the project.

The State Enterprise/OSC team's feasibility report submitted to the national government concluded that the steel project would cost U.S. $700 million, including pre-operating expenses, initial working capital, price escalations, and financial charges. Taking into consideration that all civil works, erecting some equipment structures, and engineering orders would be

placed with national suppliers, the foreign exchange
cost could be held to about 50 percent of the total.

In 1972 imported steel could be delivered without
tariff restrictions to domestic markets for about U.S.
$188 to $200 per ton (in 1972 exchange rates and cur-
rencies). This range exceeded the average price ceil-
ings for the expected product mix of the proposed plant
by 3 to 10 percent. Projections indicated that the
proposed plant could deliver steel at an average of
U.S. $186 per ton to domestic markets. This report
concluded that the plant would be able to produce steel
profitably without protection yielding a financial rate
of 11 percent. Profitability forecasts showed losses
for the two-year start-up period but increasing profits
after 1977. The net cash generated by sales through
1985 was estimated to reach $325 million, thus provid-
ing substantial contribution to new investment and
future expansion. The economic rate of return was projected to exceed 10 percent.

How the Project Was
Engineered and Implemented

The implementation of the project began in late
1972 and was completed with only a six-month delay from
the original timetable. The implementation took place
under a scheme by which State Enterprise signed contracts with both foreign and local suppliers after international bidding. The contracts called on the suppliers to supervise the installation and assembly of
their equipment at the project site. The foreign suppliers often sub-contracted the construction work
associated with the assembly of their equipment to a
local firm, while sending their own team to do the
technical installations. The joint OSC/State Enterprise engineering team supervised the construction and
assembly work while each supplier had its own supervisors in the field for their specific portion of the
overall implementation.

A bidding and preference system was adopted which
recognized that much of the procurement would necessarily comprise large packages with single contractors,
i.e., the blast furnace, pellet plant, etc. Since no
local supplier could meet prequalification criteria
for the large packages, an incentive was given to foreign bidders (a 15 percent margin of preference) if
they used clearly identified national parts and components in their bids. In those cases where local companies could prequalify to bid directly, and not simply

as sub-suppliers, their bids were also given preference.

Despite the attempts to include local suppliers and engineering services, many of the nationally-based firms felt that they should have contributed more work to the detailed engineering and the actual construction of the plant. In 1974 the project was employing an average of 50 to 60 percent of the trained manpower of the main national process engineering and construction enterprises.

As specified in the agreement between State Enterprise and OSC, most of the project engineering work was performed in the country. In 1972 as much as 60 percent of the work was performed in-country with the percentage increasing every year. In 1975 the percentage reached over 99 percent. Most of the OSC project staff also worked at the actual construction site--more than 80 percent of their total project staff by the time the first stage of engineering was completed.

A substantial effort was made to train personnel at both the managerial and operational levels. Two courses on management of steel plants, each of eight weeks' duration, were specifically designed by OSC and given to selected managerial staff from the State Enterprise. After the eight weeks of formal training, the trainees attending these courses were sent to actual operating plants abroad to learn more about what could or could not be adapted to their own local conditions.

Upon return to their country, the most promising trainees were offered individual training programs in order to prepare them as future division chiefs and assistant division chiefs. This group of trainees was well-prepared to start up the operations at the new steel plant under guidance of the foreign experts. They should have gradually taken over full responsibility for the management and handling of their respective divisions, but the transition happened more abruptly than expected because the contract with OSC had to be terminated earlier than anticipated because of financial problems in the State Enterprise.

So successful was the transfer of skills and training carried out by the foreign engineers that the engineering staff of the State Enterprise grew to the point where they announced their intention to organize an affiliated engineering firm to render service to third parties. They had nearly 250 trained civil and electrical-mechanical engineers after the steel plant was constructed.

Major problems in the implementation, however, arose in connection with delays in the housing construction. Less than 20 percent of the housing needs had been completed by the end of 1975 when the construction of the plant facility was employing nearly 10,000 workers. Severe social problems began to be felt on account of inadequate housing, insufficient consumer goods and services including high-grade foodstuffs, and lack of recreational opportunities. Labor strife broke out several times and high absenteeism was recorded at the work site coupled with low morale and diminishing productivity. In addition to the lack of physical amenities, emotional stress was experienced by many workers. The policy of discouraging the workers and professionals from bringing their families to the work site led to the spread of prostitution and alcohol abuse. A serious increase in venereal disease contributed to the worsening of the community health situation.

The expatriate consultants also suffered from many of the common psychological problems that people experience working and living in a foreign cultural environment. Apart from being concerned for their families at home, many became restless and emotionally alienated. Outside working hours, cultural "ghettos" persisted despite the fact that OSC's project manager persuaded his team not to import symbols of luxury that might highlight the socioeconomic gap between the living standards of the expatriate and national personnel.

To overcome some of the immediate problems with expatriate morale, the OSC manager instituted a series of measures. On the work site an expatriate nurse, who was fluent in the local language, was hired to act as an intermediary between the expatriates (when they became ill) and the local medical professionals. An expatriate physician was contracted to visit the site twice a year (for a total of 20 days) to provide medical and psychological consultations to expatriate personnel. A special weekly mail service between the two countries was organized to facilitate communication between separated family members.

Much of the credit for the successful transfer of skills and the averting of even greater human suffering in the early stages of the construction must be given

to the OSC's project manager. He showed considerable
sensitivity to the problems of both expatriates and
local engineers and insight into effective and cooperative project implementation. For instance, a national
engineer had been assigned to each foreign expert with
the formal authority exercised by the former. As the
differences in age and levels of training were great,
the insecurity of the younger, less experienced nationals was manifested in their strong desire to exercise authority, thus giving rise to friction between
locals and foreigners. The OSC project manager stepped
into this situation with a remarkable display of tact
and goodwill and reopened communications between the
conflicting groups. His action led to a resolution of
the problem and the re-establishment of a genuine spirit of team work during project implementation.

The environmental protective measures were also
appropriately implemented but subsequently disregarded
when the State Enterprise took full control of the
operations. The organizational structure of the Enterprise unfortunately failed to include an environmental
unit that could monitor pollution levels, offer technical and managerial solutions to problems, and inform
top management when acceptable environmental standards
were not being maintained.

Problems with dust, sometimes mixed with carcinogenic materials, appeared in and around the plant. As
a consequence, some medical personnel were fearful that
severe respiratory diseases might develop among the
workers and nearby townspeople. Pollution also began
to threaten fish and wildlife in the river and its es-
tuary. Except for one daily measurement of the efflu-
ent from the coke oven treatments, the control of all
other effluents was ignored. Severe river pollution
also occurred when one of the large concrete tanks
used for the separation of the tar and oil from the
coke oven by-product was cracked by an earthquake
tremor.* This was left unrepaired for many months
while the remaining tank was allowed to overflow into
the river.

Another potential health hazard caused by the re-
laxation of environmental controls was the burning of

---

*One of the factors not taken into account by the
project designers was the fact that the plant site was
located only 25 kilometers from a major active earthquake fault.

benzene--one of the by-products from the coke ovens.
Benzene is highly toxic and carcinogenic. The price
and difficulty of transporting the benzene made its
marketing unattractive and hence it was burned in open
pits. The combustion was far from complete and resulted in large black clouds billowing above the plant
every afternoon, often drifting over the nearby town.
In addition to the benzene, faulty devices for the pur-
ification of oven gas led to significant emissions of
sulphur dioxide and hydrocyanic acid. Fortunately for
the local population, most of the time favorable wind
conditions prevailed and blew much of the harmful gases
out to sea.

One exception to the generally poor monitoring of
pollution by the State Enterprise was the noise monitoring program. This was well established even though
measures to reduce noise levels were seldom taken. Unfortunately, everywhere in the plant (except for the
coke ovens and the lime kilns) noise levels exceeded
87 decibels. Near the blast furnace, noise levels were
recorded at an average of 102 decibels during operations. These levels were substantially higher than the
acceptable levels set during the design of the plant
and in the negotiations with the equipment suppliers.

Questions

1. What do you suppose were the major factors (constraints and opportunities) underlying the choice of
the BF/BOF route over the DR/EF route?

2. One important aspect of this case is the success
achieved in managing the transfer of technology and the
building-up of local technological capacity. What were
the key elements that contributed to this success? What
lessons can be drawn?

3. What changes could have been included in the project design to deal effectively with the human problems that arose in the early stages of implementation?

4. What do you perceive as the positive and negative
aspects of the country's procurement policy?

5. In the project design, specific attention was
given to the problems of housing, port facilities, and
railway connections. What other aspects would have
been considered in an integrated development approach?

Case No. 20

The Choice of Technology for Textile Production: The Influence of Government Policies on the Appropriate Use of Production Factors

by
Robert Maybury,
World Bank, Washington, D.C.

This case shows how government policy can have a significant impact on the choice of technology and why such choice of technique may be inappropriate. The case relates to the preference of automatic imported textile looms over semiautomatic domestic looms even though, from an economic-social cost-benefit viewpoint, the latter may have been more appropriate.

Introduction

The choice of technology in an investment project may be crucial to that project's chances of achieving its goal. This is true not only of its immediate goal of building a plant to provide a good, say, or of strengthening a government service, but also of its wider goals such as contributing to an equitable distribution of income, to generation of employment opportunities, or to balanced regional growth. It is this broader impact of the choice of technology that permits

The main ideas in this case, prepared in March 1983, and much of the text have been adapted from Y. Rhee and L. Westphal, "A Micro, Econometric Investigation of Choice of Technology," Journal of Development Economics, 4 (1977), pp. 205-237. The case was prepared by Robert Maybury and reviewed by Larry Westphal.

us to point a finger of blame at inappropriate* technological choices, asking if they are not partly the
cause of some of the problems arising in the development efforts of many countries--for instance, the failure of the industrial sector to generate sufficient
employment at an adequate wage or to build up local
capacity in capital goods manufacture. The question
needs to be asked then: Why do entrepreneurs and government officials so often make an inappropriate choice
of technique?

To seek an answer to this question, we consider a
special research study of the "choice" behavior of
decision-makers in a developing country (Rhee and Westphal 1977). The study indicates how one particular
component of the domain of feasibility--the policies
of the government in the country of the project--exer-
cise a powerful influence on the choice of technology.
Specifically, the study examines the choices made by
profit-seeking producers of cotton textile goods. The
study shows how government policies can discriminate
against the use of appropriate technologies, a lesson
that can be generalized to other countries.

Background on Constraints,
Opportunities, and Objectives

The case study summarized here examines the
choices made by the country's cotton textile weaving
firms between imported looms and locally made looms.
During the late 1960s and early 1970s, the period
covered by the study, indigenous looms sold for less
than a third of the price of imported looms. Yet, the
imported looms were purchased in large numbers.

Although appearing to be similar, on the surface,
machines may differ in many respects, that is, they may
embody different techniques of production. As a result, similar machines often cannot produce an identical range of products. Nor are they always equally
suited to make each of the products they can produce.
The width of a loom, for instance, determines the maximum width of the cloth it can weave. In addition,
some looms have comparative advantages for producing
certain grades of cloth, distinguished by the fineness

---

*"Inappropriate" signifies here a choice other
than that having the highest social benefit-cost ratio.

of the yarn or the density of the weave.* And given
certain specifications of the product, one machine may
produce output of a higher quality than another.
Furthermore, similar machines need not have the same
input requirements for skills, power, or raw materials.
Maintenance requirements also differ, as do expected
economic lifetimes.

The differences affecting the choice of a machine
extend beyond the equipment's physical characteristics.
The terms under which machinery can be purchased from
different suppliers often are not the same. Imported
machinery, for example, can often be financed by suppliers' credits, which carry lower interest rates and
more liberal repayment terms than the medium-term
domestic currency credit needed to finance the purchase
of indigenous machinery. Of course, the purchaser who
makes the choice of imported machinery on the basis of
the more favorable rates and terms offered by an outside supplier is always putting himself at risk, for
an unfavorable shift in exchange rates later on could
wipe out his profit margins. Because of investment
licensing or credit rationing, producers may not have
equal access to different sources of financing. In
addition, market segmentation and government policies
may lead different producers to pay different prices
for ostensibly identical inputs. For example, wages
paid to the same category of labor may be more readily
available at cheaper prices for some producers than
for others. Market segmentation and differential government policies can also lead to different ex-factory
prices for the same product, depending on who produces
it and where it is produced and sold.

In general, the choice of product mix and the
choice of machine specifications are inextricably in-
tertwined, and profit maximization is not simply a
matter of selecting the cost-minimizing technique to
produce a given product. The profit motive may lead
producers to choose different machine specifications
to produce identical products. It may also lead them
to produce distinct products for which each model of
machine has particular comparative advantages. Some
producers, in turn, may select their models on grounds

---

*The fineness of woven cloth depends on the thick-
ness or count of the lengthwise yarns (warp) and cross-
wise yarns (weft); its density refers to the number of
warp and weft yarns per unit length (inch or centi-
meter).

other than profit maximization, which can also explain the use of different machines.

Under the assumption that producers seek to maximize their long-term profits or net worth, economic theory states that they will compare the benefit associated with each combination of machine and product mix against its cost in order to select the combination with the highest benefit-cost ratio, that is, they will compare the discounted present value of the expected income stream--gross receipts from sales less operating expenses and taxes--against the purchase price of the machine. To investigate how various elements influence the choice of machine, it is necessary to estimate their impact on the benefit-cost ratios associated with the available models. If these estimates are to have meaning, they must incorporate a great deal of detail about the characteristics of the complementary inputs required. Just as firms may differ in important respects--for example, in their size and managerial ability--so do the factors determining the prices each firm pays for its inputs and receives for its outputs. Impinging on these factors are such things as the organization of markets and the effects of government policy.

The detailed information needed to evaluate producers' choices were obtained through a survey, conducted in 1972, of 37 cotton-textile-weaving firms.* We shall first summarize the findings concerning the differences between imported and indigenous looms and the associated differences between firms using each. Then we shall present our estimates of benefit-cost ratios derived from the data. All the imported looms considered here are automatic; all the domestic looms are semiautomatic.** The distinction between the two types depends on whether the shuttle is changed automatically or manually when its yarn is exhausted. On all looms investigated, the shuttle was passed from side to side mechanically.

Much of the cotton fabric produced was exported. There were strong government incentives for exports,

---

*The number of looms in the sample was slightly greater than a third of the total of more than 30,000 looms in the country.

**Semiautomatic looms are often referred to elsewhere as ordinary, power, or nonautomatic power looms.

so that differences in destination were associated
with differences in prices paid and received. The
share of exports in total output was greater by far in
firms using imported automatic looms. This was partly
due to the availability of indigenous looms in widths
up to only 60 inches, whereas imported looms could be
obtained in widths up to 103 inches, and much of the
export demand was for the wider varieties of cloth.
Nonetheless, the narrower exported varieties were produced by indigenous and imported looms alike.

As long as loom widths were the same, the indigenous semiautomatic looms could produce export-grade
cloth of the same specifications and quality as that
produced on the imported automatic looms. The imported
looms did have a comparative advantage in producing
superior varieties--that is, finer, more densely woven
varieties--but comparative advantage is not the same as
absolute cost advantage. For cloth 60 inches wide or
less, then, inability to produce superior varieties was
not the reason that the share of exports was lower in
firms using indigenous looms. (The findings thus contradict the often-heard allegation that older technology, in this case embodied in semiautomatic looms,
cannot produce export grade products.)

As might be expected, less labor was required to
operate an automatic loom than a semiautomatic loom:
The difference was on the order of about one to three.
The imported automatic technology also saved on raw
materials, but the difference was small. This technology involved, however, more than eight times the out-
lay per loom for maintenance and more than 15 times
that for electrical power. Moreover, based on estimated time profiles of maintenance and production
costs, the economic life of automatic looms was substantially shorter than that of semiautomatic looms:
16 years compared with 25 years.

Automatic looms were generally used by large
firms; semiautomatic looms by smaller companies organized under single proprietors or partnerships. The
average wage paid by firms using imported automatic
looms was 85 percent more than that paid by firms using
domestic semiautomatic looms. As might be expected,
wage differences were associated with the occupational
mix of the labor as well as with the plant's location
and type. The mechanisms for funding purchases also
differed. Nearly all imports of automatic looms were
financed by foreign suppliers' credits on which the
real interest rate was under 2 percent. Most purchases

303

of semiautomatic looms, on the other hand, were financed by rolling over short-term domestic credit on
which the real interest was 14 percent.

There were also important differences in tax
treatment, depending on the share of output exported.
For example, there was no indirect tax on yarn used to
produce for export, but there was a minimum tax rate of
10 percent on yarn going into production for domestic
sale. The business activity tax (a turnover tax
charged at a maximum rate of 0.5 percent of gross
sales) was not levied on exports or on inputs used to
produce exports. The schedule of income tax rates was
progressive, with lower rates on corporate income. More
to the point, income derived from export sales was
taxed at half the rate otherwise applicable. Furthermore, exporters were allowed a somewhat higher rate of
depreciation in calculating their business expenses.

The country's cotton-textile-weaving sector at the
time of the survey might be characterized as having a
dual structure. Firms paying high wages and benefiting
from low interest rates typically used the more expensive, automatic looms and tended to produce superior
varieties, mostly for export. Firms paying low wages
and high interest rates characteristically used the
less expensive, semiautomatic looms and tended to produce inferior varieties, mostly for the domestic market. This pattern is consistent with the notion that
differences in economic environments led to the selection of different technologies. But a rigorous test
of this notion requires the estimation and comparison
of benefit-cost ratios for the alternative technologies.

Technological Relationships

The estimation of the benefit-cost ratios for each
of the two technologies makes use of detailed information gathered from firms at the man-machine level:
Characteristics of machine models and of the goods produced; characteristics of the required complementary
inputs; size and management capabilities of firms;
etc. These data permitted estimation of technological
relationships for different machines, relationships
that distinguished among loom, fabric, and complementary input specifications as well as firm characteristics. These relationships lie at the heart of the
benefit-cost ratio determinations. We illustrate the
approach followed by the use of one such relationship
--that which determines the number of looms tended by

one operator (a sort of capital-labor ratio).

This relationship that "governs" the number of looms tended by a single operator depends upon a number of factors, including the characteristics of the product being made, the machine being used, the firm doing the production, as well as the skill levels of the labor being employed. In addition, productivity changes over time because learning-by-doing and/or physical deterioration of the machine are also included in the relationship. Each variable entering the equation for this relationship is briefly described below.*

Product characteristics

The most important product characteristic in the case of both technologies is the fineness of the yarn being woven; a higher count indicates a finer yarn and, generally, a superior fabric. There is a major difference between the semiautomatic and automatic technologies with respect to variations in the fineness of the yarn used. With the semiautomatic technology, the required labor input rises (i.e., the number of looms per operator falls) with a rise in yarn fineness. With the automatic technology, it is just the reverse--the required labor input falls (i.e., the number of looms per operator increases) as fineness improves.

Loom characteristics

Thus, the operation of a given number of domestic semiautomatic looms requires more labor than does the operation of the same number of imported automatic looms, assuming the values of the other independent variables to be constant.

Semiautomatic technology. Under this technology, the shuttle is changed manually. The volume of the shuttle (length times width times height) determines how much yarn it can carry and thus how often it must be changed. As one would expect, the use of larger shuttles reduces the requirement for operating labor. A "feeler" is a simple device which indicates that the yarn in the shuttle is almost exhausted and thereby obviates the need for the operator manually to check the yarn supply in the shuttle. With feelers, the number of looms that can be tended by a single operator

---

*Although not presented here, specific parameter values were estimated for each variable.

is increased by roughly 10 percent.

   Automatic technology. The size of the shuttle is
not significant under the automatic technology. Rather,
the width of the loom and the type of mechanism used to
replace or reload the exhausted shuttle are the most
important factors. A wider loom requires more time to
be set up to produce a particular variety of cloth and
thus requires more operating labor. In turn, there are
three types of mechanisms for replacing or reloading
the shuttle. In the first of these, the exhausted
shuttle is simply replaced by another, fully loaded
shuttle. The other two do not change the shuttle but
instead simply replace the "cop," the component of the
shuttle that actually carries the yarn. Changing the
cop saves labor relative to changing the entire
shuttle, with a magazine-type mechanism for changing
the cop being labor-saving relative to a box loader.

Firm characteristics

   The data set includes four firm characteristics:
Size, location, type of ownership, and type of office.
Firm characteristics appear to make a significant difference only under the semiautomatic technology. While
firm size and type of office are highly correlated
among firms using this technology, the latter characteristic appears to be the more significant. Thus,
firms having a residential office and which may be considered to be in the informal sector require roughly 17
percent more labor to operate the same number of semiautomatic looms. However, this is not sufficient evidence to conclude that these firms are inefficient. In
fact, their overhead expenses are less than those of
firms having nonresidential offices and they exhibit
significantly lower material wastage rates.

Wage rates

   The wage rates paid to loom operators may influence the number of looms tended per operator for at
least two reasons: Higher wages may reflect higher
skills, or if the skills are equal the producer may re-
spond to higher wages by substituting other factors for
labor. For example, when confronted with a higher
efficiency wage, the producer might substitute raw
materials for labor despite a higher wastage rate when
each operator tends a larger number of looms. Here one
would definitely expect the loom operator's wage also
to be a significant determinant of the use of the in-
put(s) with which the loom operator's labor is

306

substitutable.

In the case of the semiautomatic technology, wages were found to have a significant positive influence on the number of looms tended per operator. Both of these explanations seem to apply. Wage rates seem not to influence the number of automatic looms tended per operator.

Rise in productivity

Certain variables and corresponding explanations were tested for their significance in determining rates of increase in looms tended per operator over a period of time. Among these were the loom's age--to capture the loom's physical deterioration with use and learning-by-doing effects at the loom level; the loom's vintage--embodied technological change; the firm's age--learning-by-doing at the firm level; and, simple calendar time--disembodied technological change.

The number of looms per operator approaches a constant, asymptotic value under the automatic technology; nearly 80 percent of the increase is realized within the first five years of the loom's operations, during which the average annual rate of "labor productivity" growth is 4.8 percent.

The average annual rate of "labor productivity" growth during the first five years of a semiautomatic loom's operation, 4.7 percent, is nearly equal to that for an automatic loom.

While the evidence points to an extremely rapid rise in productivity during the sample period, it is unrealistic to extrapolate this too far into the future. Textile engineers commonly assume an upper limit to the number of looms per operator which it is physically impossible to exceed.

Other technological relationships

As stated previously, the technological relationship just discussed is but one of several required to estimate the benefit-cost ratio of a particular loom. Three additional categories of labor must be distinguished: Cop suppliers, who assist the loom operator by refilling the exhausted cop; loom engineers, who are responsible for scheduled and unscheduled maintenance; and assistant loom engineers. For each of these, a technological relationship governing the number of

307

looms serviced by one laborer was estimated.

Yarn wastage and fabric rejection rates were estimated using individual technological relationships, as was the efficiency ratio, which is the fraction of time that the loom is actually weaving as opposed to being idle for resupplying yarn or repairing yarn breakages. These relationships, plus several engineering identities, determine the amount of yarn wastage, the quantity of rejected product, the gross yarn requirement, and the rate of fabric production. The equations governing yarn inputs and fabric outputs include as explanatory variables the characteristics of the fabric being produced.

Additional technological relationships are used to determine the amount of power consumption, the cost of maintenance materials, the cost of yarn preparation, and overhead cost.

Evaluation of Loom Choices

For each model of loom in the sample, several benefit-cost ratios were estimated, one for each principal variety of fabric produced on that loom.* The number of loom models using each technology and the number of loom and fabric combinations for which benefit-cost ratios were estimated are shown in the table, which summarizes the estimates.

The estimated "private" benefit-cost ratios are based on prices that producers paid and received. The estimates thus reflect all the differences in circumstances associated with the purchase and use of different looms, including those attributable to government price policies. They also incorporate productivity changes from the date of a loom's purchase to the estimated date of its scrapping. And they embody reasonable forecasts of changes in relative prices--such as rising real wages--just as they factor in observed changes over the period following the purchase of the looms.

Although the average private benefit-cost ratio is different for the two technologies, the difference is not statistically significant. On average, no

---

*Differences between loom models reflect differences in suppliers, vintages, and specific characteristics.

Benefit-Cost Evaluation of
Alternative Weaving Technologies

Items	Domestic Semiautomatic Looms	Imported Automatic Looms
Sample size		
Loom models	26	32
Loom-fabric combinations	65	73
Average benefit-cost ratio		
Private	4.14	4.39
Social	5.48	1.49
Average subsidy rates		
Percentage increase in benefit due to policies affecting:		
Interest rates	-67.8	76.8
Direct and indirect taxes	-36.5	19.4
Input and output prices	128.8	53.7
Percentage decrease in cost owing to policies affecting loom purchase price	0.0	21.3

Note: Averages are simple averages over all the
loom-fabric combinations for which estimates were made
under each technology.

producer could have realized higher profits by changing
either his circumstances or his choices to make them
more like those of any other producer. In this respect, the estimates are consistent with the hypothesis
that the producers chose loom models on the basis of
profit maximization. That is, it appears that differences in prices paid and received, as well as in the
varieties produced, led to the choice of different
technologies.

The Effects of Government Policies

We now turn to evaluate the technologies at uniform prices.* Our purpose in doing so is to remove distortions caused by government price policies. The procedure for estimating "social" benefit-cost ratios is the same as that used to obtain estimates in the private case. The loom-fabric combinations for which benefit-cost ratios were estimated are also the same. But the following changes were introduced to impose uniform prices:

● To remove the effect of differences in interest rates, all loom purchases were assumed to have been financed on identical terms, i.e., equal to the average of the observed terms.**

● To remove the effect of differences in direct and indirect tax treatment, all indirect tax rates were set at zero, the average actual direct tax rate was imposed uniformly, and depreciation deductions were calculated everywhere on the same basis.

● To remove the effect of trade policies that protected domestic producers selling in the local market by affecting the prices of intermediate inputs and of outputs, the prices of all varieties of yarn and

---

*The imposition of uniform prices is akin to using shadow prices to calculate the benefit-cost ratios. Overvaluation of the currency is taken into account by using the appropriate shadow exchange rate, while tradable goods are valued at border prices. But labor is not shadow-priced and the uniform discount rate is simply taken to be the average of the real interest rates paid to finance loom purchases in the sample. On the other hand, the market wages in the country under study were not different from shadow wages, so that it is principally in underestimating the discount rate that the approach taken in this study markedly differs from the use of shadow prices.

**Through payments to banks, the government explicitly subsidized domestic credit made available on preferential terms. On the other hand, government licensing merely controlled access to foreign suppliers' credits. Nonetheless, with respect to the latter, the government could have imposed an interest equalization tax but did not, so that access to the lower interest rate must be considered as an implicit subsidy.

fabric were set equal to their respective world prices converted at an appropriate exchange rate, and tariff rates were set at zero.

● To remove the effect of an inappropriate exchange rate that artificially reduced the cost of imported machinery, the purchase prices of imported looms were increased in proportion to the degree of exchange-rate overvaluation.

It was assumed that there were no differences in treatment of production for the domestic and export markets and that fabrics were sold domestically at the same prices prevailing in the world market. The resulting estimates are the social benefit-cost ratios shown in the table.

Social benefit-cost ratios exhibit far greater differences than do private benefit-cost ratios. Moreover, the average social benefit-cost ratio for the domestic semiautomatic technology is much greater than that for the imported automatic technology, and statistically the difference is highly significant. It therefore appears that few, if any, imported automatic looms would have been purchased in the absence of preferential government policy, except where required to produce fabrics wider than could be produced using indigenous looms. But of 89 directly observed loom-fabric combinations using the imported automatic technology, in only 22 did the fabric widths exceed the production capability of indigenous looms.

It can be concluded that the indigenous semiautomatic technology was the socially optimal choice to produce fabrics less than 60 inches wide.* For the same initial investment, use of the domestic semiautomatic technology would have generated more than 10 times the jobs associated with the imported automatic technology. In addition, less skill is required to operate the semiautomatic looms, and their economic life is substantially longer.

By estimating benefit-cost ratios under alternative sets of assumptions, it is possible to break down the difference between the private and social benefit-

_____

*Extensive sensitivity analysis showed that the semiautomatic technology retains its superiority over the automatic technology in a wide range of circumstances.

cost ratios for a particular loom-fabric combination into components attributable to different policy elements. As with any such analysis, there is a certain degree of arbitrariness in determining the separate contribution of each element, but the results are nonetheless instructive. The table shows the average subsidy rates associated with removing each of the distortions in going from private to social benefit-cost ratios. These rates indicate the percentages by which the social benefits and costs were increased or decreased as a result of government policies affecting the indicated prices.

Not all the subsidies considered here were explicitly tied to the purchase of technology. Nonetheless, those related to the purchase of imported automatic looms--that is, access to preferential credit and the overvalued exchange rate--on average were sufficient to increase its benefit-cost ratio 125 percent.* Subsidies originating in tax differences and import protection, the latter affecting prices paid for inputs and received for outputs, were not tied to particular technologies but were realized through producers' choices of the varieties of fabric to produce and the markets in which to sell them. On average, all producers benefited from protection in the domestic market, but only those using the imported automatic technology benefited from preferential tax rates associated with concentrating their sales in export markets.

Conclusions

Without estimates of demand relations for different fabrics, one cannot analyze the quantities of various fabrics that would have been produced under different circumstances--analysis needed for closing the circle between product demand and technology choice. Nonetheless, the foregoing discussion shows that government incentive policies were not neutral. In particular, export incentives in combination with cross subsidization between domestic and export markets appear to have elicited a larger volume of production of superior fabric varieties than would otherwise have been so. (Higher prices on the domestic market, sanctioned by measures to protect domestic products in that market, enabled lower prices on the export market.) In

---

*Refer to the figures shown in the table and note that [(1.0 + 0.768) / (1.0 - .213) - 1.0] yields 124.7 percent.

turn, the choice of the imported automatic technology was generally associated with the production and export of these superior varieties. It bears emphasizing that one effect of the underlying policies was to increase prices to domestic consumers because some producers chose socially inappropriate techniques, which meant higher production costs though not higher profits.

To summarize: Producers' choices generally seem to have been consistent with profit-maximizing behavior. That the choices were also socially inappropriate is explained by the constellation of government policies favoring the use of imported technology. Particularly important among these policies were access to credit on preferential terms and exemption of machinery imports from tariffs, together with an overvalued exchange rate. Except for the production of cloth more than 60 inches wide, the indigenous semiautomatic technology would have been the socially optimal choice of technology. And beyond the qualification just noted, there is every reason to expect that producers would have chosen this technology under a more appropriate set of policies. The policies encouraging the use of imported technology simultaneously discriminated against domestic textile-machinery manufacturers and thus inappropriately retarded the development of the domestic engineering industry.

Since the time of the study, the country's government policy has changed. In particular, imports of capital goods are no longer favored through tariff exemptions or access to credit on softer terms. In some cases, there are even import restrictions on competing imports of equipment. Income derived from exporting is no longer taxed at a preferential rate, and other export incentives have been substantially reduced. Many of the differential policies have thus been changed or removed. And it appears that export prices for most fabrics now are equal to or higher than domestic prices, even when allowance is made for the remaining export subsidies. As a result of the engineering industry's increasing maturity, wider looms are now produced domestically so that the imported technology no longer has an absolute advantage on this score. Equally significant is the fact that more than two-thirds of the new looms purchased by cotton-textile producers in 1973 and 1974 (the latest years for which aggregate data were obtained) were locally manufactured.

313

Questions

1. In their effort to attain the highest benefit-cost ratio for their firms' performance, these profit-seeking producers are responsive to price signals. What are the various items carrying a price that enter into the concern of these producers?

2. For each of these items in Question 1, review how specific government policies (taxes, interest rates, etc.) will influence the price it carries.

3. Where the influence on price of a policy is socially adverse, discuss the change required in the policy to eliminate this negative influence.

4. Describe the situation surrounding each type of producer--large firm and small entrepreneur--and discuss how this situation affects the preference of each for the kind of technology chosen as well as for the particular product mix produced.

5. Describe the "social" losses felt by the country's economy due to (1) the "inappropriate" choice of looms by the large firms, and (2) the government's policies.

Case No. 21

Youth Brigades:
Technology Choice in an Education Project

by
Paul Bundick,
Word Bank, Washington, D.C.

This case highlights the creation of youth brigades in a developing country as an educational alternative to promote employment, production, and ultimately economic development. Especially, the case
stresses the human values and cultural orientations of
the project designers in choosing among various technological options.

Introduction

Youth unemployment is a chronic and pervasive
problem in many parts of the world. Owing to a number
of factors such as increasing population, inadequate
education, and lack of capital for productive investments, many countries have not been able to generate
enough wage-paying jobs to absorb the growing numbers
of young people seeking entry into the labor force.
These unemployed youth frequently include those who
have graduated from the primary schools in rural areas.
Even though they are literate, these primary school-
leavers often lack skills which can be translated into
wage employment either in the cities or in their own
rural environment.

In the past, many developing countries have in-
herited educational systems or designed new ones based

This case was prepared by Paul Bundick in October
1982 using World Bank documents and additional information provided by Nat Colleta (AEA), who critically
reviewed preliminary drafts.

315

on models from industrialized countries. These approaches frequently have been ill-suited to solving the unique problems of each particular nation. The curriculum has often tended to emphasize theoretical studies while insufficient time has been devoted to practical training or attempts to link education with employment opportunities.

This case looks at an educational alternative in a developing country that was created by a non-governmental organization to promote rural development by linking education with the generation of employment. The approach connects education with production and goes well beyond traditional classroom methods or the conventional vocational training centers. The youth brigades, as they are called, strive to create new job opportunities and recover their costs of operation through the sale of products and services. They seek not only to develop skills through the process of productive labor but also try to encourage new values and attitudes which are conducive to the development of rural areas.

In the case study, the Brigade movement was identified by the country's Ministry of Education as an educational technology which could be adapted to solve the serious youth unemployment problems as well as help alleviate some of the manpower shortages in the modern sector. In an attempt to improve the educational system, the government decided to expand the Brigade movement not primarily for the purpose of rural development but to train rural youth for wage employment in the urban areas. Interesting issues are raised regarding the theory and practice of education linked with production and the conditions under which it can be used effectively. Also highlighted in the case is the importance of the philosophical and value orientations of the project designers in the selection of an overall technology. Values are key elements in determining the range for the domain of feasibility when choosing among various technological options.

The Setting

The project under consideration was carried out in a relatively large developing country of more than a half million square kilometers. Despite its considerable size, most of the country is desert, virtually uninhabited except for nomadic tribes. Only limited areas have sufficient rainfall to permit agricultural settlements. Most of the country's 700,000 people are concentrated in these areas of higher rainfall, a narrow strip of savanna where precipitation averages 500 millimeters a year. There, the majority of the people earn a meager livelihood through subsistence farming and livestock production.

For centuries, agriculture and livestock raising have been inextricably linked for the majority of rural people. In the savanna grasslands, periodic drought, primitive cultivation techniques, and generally poor soils have made agriculture a risky undertaking. In this unpredictable environment, cattle have given the needed insurance against crop failure. In good years, cereal crops provided the major source of food, while the herds were conserved and built up. In times of drought, meat substituted for grain as the staple food. Livestock consumed crop residues and grazed on non-arable land and also provided draft power for cultivation and transport.

In the traditional society, labor was highly specialized by age and gender. Young boys did the herding while young girls scared birds away from the crops; adult men hunted, butchered the cattle, and took responsibility for the herds while women did the crop planting and harvesting. Local industries were often left to the elderly. Carpentry, iron-work, and leather tanning were the domain of the males. Females made pottery, brewed beer, and fashioned beadwork. Because of their association with cattle, men ploughed the fields although women have always done the major part of the agricultural labor.

In pre-colonial times the society was made up of self-sufficient villages based on tribal bonds. Barter and reciprocities ("my oxen at ploughing time for your labor now") substituted for money. Wealth was stored in the form of cattle. As cattle had no income-producing capacity in the village economy, the standard of living of both the "rich" and the poor were very much the same. All prospered when the weather was propitious; all suffered when it was not. The traditional social structure, however, could not be described as egalitarian or communal. Abundance was not so much shared as exchanged for obligations. Authority was the prerogative of the village patriarchs who made up the governing hierarchy. One of the duties of the village leaders, however, was to insure that each household had adequate community support in times of need.

In recent years, the traditional society has undergone rapid changes. Since the 1960s the country's Gross Domestic Product (GDP) grew at an impressive rate of 20 percent per annum, and per capita income increased fivefold (from U.S. $60 to U.S. $300 per annum). This rapid rate of growth was primarily due to the government's efforts to build up the modern sector coupled with the discovery of extensive mineral deposits and the growing export of meat. This led to the creation of a separate cash-fueled economy and a dynamic formal sector which altered many of the traditional social patterns and economic relationships.

The expanding commercialization of the economy did much to break down traditional authority, increase the gap between rich and poor, and contribute to massive rural to urban migration. Government officials and administrative policies did much to erode the organizing powers of the tribal leaders and hence their ability to mobilize their villages for joint enterprises and even to provide for the poor. Many of the young men between the ages of 18 and 35 left the plough for the mine and the factory. Cash remittances sent home by workers became an important part of the village economy. During the 1960s and 1970s the increasing monetization of the economy resulted in a greater need for money to survive and the selling of commodities which were traditionally traded or given away. With the improved infrastructure and marketing opportunities, the large cattle owners converted "rural wealth" into rural income. People who owned no cattle became poorer and often destitute.

During the mid-1970s, the livestock industry accounted for more than one-third of the GDP with the rapidly growing mining industries representing over 15 percent of the domestic output. Minerals and cattle products provided nearly all the nation's exports and foreign exchange earnings needed to meet the increasing demand for imported goods. In this period the country's agricultural sector (apart from livestock) did poorly and failed to keep pace with rising demand. From a position of food self-sufficiency in the 1940s and 1950s, the country became a net importer of food in the 1970s. In the early 1970s, the average yearly crop yields were only sufficient to supply two-thirds of the domestic need. Besides food and beverage products, other imports consisted of manufactured goods, chemicals, fuel, and machinery required by the modern sector. From 1970 to 1974, imports (as a percentage of the GDP) increased from 50 to 60 percent while exports rose from 30 to 40 percent.

The rapid economic growth and the relative rise in the importance of the modern sector resulted in a significant increase in wage employment. In 1968 there were only about 25,000 people working for wages in the country. In 1974 the number reached 50,000 (excluding 8,000 domestic servants) and comprised 16 percent of the active labor force. The important structural changes in the economy were also reflected in the sectoral distribution of employment. While the share of agricultural wage employment declined, those of construction and mining increased. Commerce, finance, services, and government jobs together comprised more than half of the country's total wage employment, as shown in Table A below.

Table A

Percentage of Wage Employment by Sector

	1968	1974
Commercial/finance/services	27%	23%
Central government	24	24
Agriculture	17	11
Education	7	7
Manufacturing/electricity/water	7	6
Construction	6	14
Transport	5	3
Local government	4	4
Mining and quarrying	3	8
	100%	100%

The lure of jobs in the commercial areas and the mining centers led to massive labor migration both inside the country and to foreign countries. The rapid internal migration was reflected in the high urbanization rate. In the 1960s and 1970s, urban population increased at 15 percent per annum. Between 1971 and 1974 the population in the capital city more than doubled. The migration to foreign countries was even more significant. In 1971 an estimated 46,000 people were working in one neighboring country alone. The existence of such a large expatriate labor force was due to liberal trade agreements between the two governments and the fact that the minimum working wage in the foreign country was twice the domestic rate.

The trade agreements between the two countries allowed for a virtual free exchange of goods. During the 1970s, over 50 percent of all the nation's imports originated from this nearby foreign country which had a stronger industrial base. Although the arrangement brought increased revenue (as much as 12 percent of the GDP), labor remittances and product availability, domestic producers in agriculture and industry were often unable to compete with the quality and price of the imported goods because of the more efficient production techniques of their economically more powerful neighbor.

In accordance with the trade agreements, the government required any new business to demonstrate that it could supply the entire national market before it could claim protective tariffs. This was normally only possible in the case of large-scale industrial developments heavily financed by foreign capital. As a consequence, most rural enterprises remained small while supplying only a limited market, and they were always vulnerable to any influx of imported goods. During the 1960s and 1970s, many urban and foreign-based industries successfully penetrated the rural areas with their products and put hundreds of small producers out of business.

The country's rapid economic growth derived from capital intensive investment in mining, and modern industries primarily benefited the urban people, or about 15 percent of the population. The urban-rural income disparity in 1974 was estimated at four to one. There was also a high income inequality among the sectors of the economy. In 1974 average monthly earnings of central government employees were twice those of the total population employed for wages. Also, the earnings of government employees were six times greater than those of the wage-earners in the agricultural sector. The government's minimum wage, designed to prevent the exploitation of urban workers in private industry, caused many skilled and educated men in the rural areas to migrate to the industrial and mining centers. Rural enterprises could seldom compete with the relatively high salaries in the modern sector.

Income disparities reached even greater inequalities in the rural areas. The large cattle owners benefited greatly from the improvements in infrastructure and increases in meat exports. New roads and slaughterhouses allow them to convert much of their stocks into cash incomes. In the mid-1970s about 5 percent of the

households in rural areas owned more than 50 percent
of the nation's total cattle herd while nearly one-
fourth of the rural households owned no cattle at all.
In 1974 the poorest 5 percent of the rural households
(averaging six persons) had yearly incomes of less than
$220 taking into account both cash income and subsist-
ence activities. At the other end of the spectrum, the
wealthiest 5 percent of the rural households all had
yearly incomes exceeding $3,700 with a few families
making more than $60,000. The majority of the people
in the rural areas did not benefit from the remarkable
economic advance; many rural incomes actually declined.
By 1974 one-half of the population in the rural areas
lived in real poverty with annual incomes of less than
$130. Particularly disadvantaged were the 27,000 rural
households headed by women.

     Recognizing the growing gap between the rich and
poor and the urban and rural areas, the stated overall
development policy of the government emphasized the
need to extend the benefits of development to all citizens. The development plan for the later part of the
1970s outlined four major goals: (a) Rapid economic
growth to provide a means to improve social welfare;
(b) equitable distribution of income particularly reducing the income differential between urban and rural
sectors; (c) sustained production through conservation
and careful use of natural resources; and (d) economic
independence and self-sufficiency. The development
plan called for capital-intensive investment in the
mining industries, investments in agriculture and rural
infrastructure, and labor-intensive enterprises to
create employment and education to provide a skilled-
labor force.

Education

     In 1973 a government manpower survey revealed that
the educational background of the country's wage-
earners and professionals was among the lowest in the
world. Only 10 percent of the nationals in professional and technical jobs had a university education and
over 60 percent had left school after junior secondary
level. The study also discovered that 82 percent of
all professional positions in the country were held by
expatriates. Foreigners also accounted for half of
the workers with senior secondary schooling and 20
percent of those with junior secondary education. The
low educational background and the lack of skills among
the native population were becoming major constraints
to economic growth and self-sufficiency.

The shortage of trained manpower caused the government to create the Directorate of Personnel, a kind of national employment service, in 1974. The Directorate was responsible for recruiting skilled nationals and allocating them to government and private employers. Records indicated that in 1974 the Directorate could meet only 55 percent of the requests from potential employers for junior secondary school graduates and only 17 percent of the requests for senior secondary school graduates. As the country continued to grow, the gap widened between the demand for and the supply of trained and educated manpower.

Prior to 1966 the country had never developed a public school system. At that time there were fewer than 200 private primary schools and only four secondary schools operated by religious missionary groups. Although the system had expanded greatly by the mid-1970s, the sector was still unable to produce enough graduates to meet the manpower requirements of the modern sector. In addition, the general school curriculum was overly literary and included little which would prepare students to earn a living on a farm or in the wage economy.

By 1974 the educational system covered primary, secondary, and university education. It also had components for technical and vocational training, teacher training, and nonformal education. The Ministry of Education had overall responsibility for all educational activities, except for training activities such as agricultural extension services which were organized by the different respective ministries.

The primary education course covered seven years and led to the Primary School Leaving Certificate. Since 1971, enrollments in primary schools increased by 60 percent. By 1974, 116,200 students were enrolled in primary schools, 89 percent of the age group 7 to 13. Unlike the primary schools, secondary schools were divided into two cycles: Junior secondary school (grades 8 through 10) which required three years to complete and culminated in the Junior Certificate Examination, and Senior Secondary School (grades 11 through 12) which took two years and ended with a Secondary School Certificate. In the early 1970s only 18 percent of the primary graduates entered the secondary cycle. Because of the manpower shortages in the modern sector, the government set a policy that 50 percent of the secondary school students had to leave school after receiving their Junior Certificate. Thus, only half of the junior secondary school-leavers had the opportunity to continue through the senior secondary cycle. Statistics indicate that only about half of those students fortunate enough to continue their education beyond the Junior Certificate level ever complete their schooling and earn their final Secondary School Certificate.

In general, primary and secondary education were designed to prepare students for the next level in the educational system with little attention devoted to practical studies in agriculture, handicrafts, domestic science, or engineering. The formal educational system provided little support for the traditional sector of the economy or the large majority of primary school-leavers who did not have access to secondary education or specialized technical training.

Several government and private agencies by the 1970s provided formal vocational and technical training. The National Training Center offered training in the general engineering trades. The government's National Center for Vocational Training gave courses at the craftsman/technical level in carpentry, plumbing, electrical skills, mechanics, and welding. The facilities were expanded to focus on middle and higher level technical training. Other institutions emphasized clerical and administrative skills needed in both government offices and private industry. In general, however, the expansion of formal vocational and technical training had not kept pace with demand. In 1974 the enrollment at the National Center for Vocational Training was only 155. Shortages of manpower and finances impeded adequate expansion of vocational education facilities.

Youth Brigades

One of the nonformal skill training programs created within this educational context was the Youth Brigade. The first Brigade was founded in the mid-1960s by a school principal who felt that the curriculum of the traditional academic school system was not meeting the students' needs regarding vocational skill training or employment, especially the estimated 70 percent of primary school-leavers unable to attend school beyond grade seven. Large numbers of primary school graduates lacked skills for wage sector employment and the rural areas could provide jobs for only a small percentage of the unemployed youth. The first Brigade consisted of a group of primary school leavers being trained by a qualified builder in the

323

construction trade.

The most important feature of the Brigade movement was its linkage of training and production. From the movement's inception, the founder viewed training and production as two necessary and complementary functions, each designed to serve the other. Skill-training was provided on-the-job, and the Brigade members and work supervisors were also instructors and teachers. Production was geared to both training and education. Trainees (who were also workers) learned skills through doing productive work which in turn helped pay for the training costs. The movement also stressed motivation and values and the need for a mass-based pedagogy which would raise the cultural level of the rural areas and give people the tools to understand and control the political, economic, and social forces which determined their lives.

The original movement set down the following principles as guidelines for their training model:

1. Vocational training shall be provided mainly for the primary school-leavers outside the formal education system.*

2. Training shall be primarily geared to the needs of the local area. The general aim is rural development to be achieved by offering training for gainful employment.

Generally the government's experiments proved less successful than the community-initiated Brigades and by the early 1970s most Brigades were owned and operated by local trusts. Local trusts varied in organizational structure but most had features in common. Both the original movement and most of the second-generation experiments were legally autonomous units incorporated as non-governmental organizations.

In most cases the term "Brigade" was used to emphasize the importance of production and to differentiate its methods from traditional vocational schools. The original Brigades devoted approximately 80 percent of their training time to actual production. In the remaining 20 percent of the time, trainees were given theoretical instruction related to the skills being

---

*Formal includes the traditional vocational and technical training.

learned, such as carpentry or construction theory. Mathematics, language, science, and rural development studies were also taught in the classroom up to four hours a week.

In the Brigade training, theory was always reinforced with practical applications in the workshop. All operations were first demonstrated in a non-production setting. Then new trainees were placed to work with experienced trainees, thus allowing continued learning from more experienced student workers. As trainees became more proficient in a skill, they were allowed greater responsibility in the production process. In order to develop leadership and a sense of responsibility, trainees were encouraged to participate in decision-making within the Brigades as well as take an active part in the social and economic life of the local communities. Enrollment in the Brigades was voluntary with training being given to both boys and girls. Importance was given to the adoption of "development-oriented values" and the "Brigade ethic." This ethic is difficult to define but primarily refers to the idealism, high motivation, and local initiative which were part of the ideological foundations of the early Brigade movement.

The founders of the early Brigades saw the movement not merely as a skill-training program but rather as something far more fundamental--a means to transform the social and economic life of the rural areas. They argued that education was never neutral but either perpetuated socioeconomic inequalities and dependence on elites or contributed to the transformation of the economy to meet the needs of the vast majority of the underprivileged. They felt that only by linking education with production and thereby contributing to the solution of economic underdevelopment in the rural areas could universal education be made financially possible for all.

Self-sufficiency and cost recovery operation were, therefore, important values in the Brigade movement even though only a few centers were able to consistently cover their costs. Brigade units often operated at a loss and were subsidized from within the Center by another "profit-making" Brigade. Generally, agricultural Brigades did poorly while those Brigades emphasizing trade skills fared much better. While financial self-sufficiency continued to be stressed throughout the movement, most Brigade Centers only recovered between 50 to 90 percent of their operating budget.

Poor business management, salary demands to keep pace with modern sector wages, and difficulties in marketing all contributed to the Brigade's problems in attaining complete self-sufficiency. External assistance agencies, private donors, and the national government all contributed at times to the movement, but still the largest portion of Brigade income was generated by the sale of goods and services which the Brigade produced, whether it was clothing, construction services, or farm products.

Linking education with production, however, was not merely a financial expedient to the Brigade philosophy. The movement originally saw linking learning with productive work as the first step towards bringing school closer to the social and economic reality and overcoming the gap between theory and practice. Education was viewed as not just a passive acquisition of skills and knowledge in a classroom or trade school but rather as an active combination of mental, emotional, and physical activity with "real life" involvement. Many stressed the need for changing attitudes in the educational process and fostering values of cooperation, responsibility, social justice, and the economic betterment of their local community. Since its founding, the Brigade movement has made steady progress in training skilled manpower for employment at a basic vocational level.

The Project and Technology Choice

But in spite of the steady expansion of the Brigade movement in the late 1960s and early 1970s, the problem of youth unemployment in the rural areas did not substantially improve. The nation's rapid population growth rate (nearly 3½ percent) continually filled the ranks of the unemployed youth. By the mid-1970s more than half of the country's population (exclusive of migrant laborers) was under the age of 15 years.

In 1975 there were 66,000 primary school graduates in the 14- to 18-year-old age group. Of this total, 12,000 were enrolled in secondary or junior secondary school full-time and 6,000 were employed either as skilled or unskilled workers in the modern sector. The remaining 48,000 young people were either unemployed or engaged in the non-monetized informal sector. They constituted more than 70 percent of this age group and were not likely to find wage employment without additional skill training. Within this context, the government decided to design an investment project in vocational training that would help alleviate the youth unemployment and train several thousand young people to take jobs in the modern sector.

The Brigade training model was generally recognized by the government for its contributions in the area of low-cost vocational training. However, the Ministry of Education claimed that, with a few notable exceptions, the Brigade vocational training had not reached acceptable standards when judged against the national training system. Critics charged that the principal reason for this lack of consistent high-level standards was the Brigades' insistence on recovering their costs through production. The Ministry of Education felt that by emphasizing production instead of skill acquisition, the quality of training suffered. It argued that one day per week spent on theoretical learning was insufficient for the mastery of cognitive skills essential to meet the high standards required by the modern sector.

The government also noted that the Brigade model was primarily designed to train youth for rural employment although many Brigade trainees had found jobs in the urban centers. While many of the original Brigades had used "through-put" training methods,* the increasing emphasis in the movement was "stay-put" training coupled with the aim of generating productive enterprises and jobs in the rural areas. The Ministry of Education stressed the need for "through-put" methods for its vocational education objectives.

Since the movement's inception, the Brigades in general were regarded with a great deal of suspicion. The emphasis upon social and economic change and the "Brigade ethic" threatened some government officials. Others saw the emphasis upon "new values" of cooperation and participation as unworkable in a vocational training school which required the grading of students and the delivering of trade tests to determine competence levels. Some felt that individual incentives

---

*Through-put refers to training designed to put skilled persons into the labor force. The trainee passes through the training institutions and migrates to where the jobs are. Stay-put training, by contrast, trains people for employment at the job site. It is designed to prepare people in their environment for local employment.

should be introduced along with standardized tests to ensure high quality training in the Brigade Program.

In the 1970s the Ministry of Education began to offer services to individual Brigade Centers to upgrade the level of training. The assistance included formal courses in management, bookkeeping, and teacher training. The government also began to recruit overseas volunteers and provide the Brigades with information and supplies. These services were accompanied by additional measures to integrate the Brigade Movement into the national educational system.

Brigades were only given grants and services if they met a series of conditions established by the Ministry of Education. In order to receive a government subsidy, a Brigade had to "truly" represent the interests of the community in the area of operations, keep financial records in accordance with Ministry guidelines, maintain acceptable standards of throughput vocational skill training, and allow for government representation on its Board of Trustees. Student subsidies were also given to only "approved" Brigades involved in purely through-put training. The aim of the subsidy was to ease the financial burden imposed by the cost-recovery production so time devoted to "actual" skill-training and instruction could be increased. Subsidies were designed to allow Brigades to spend less on production and hire better qualified staff.

By the mid-1970s as an outcome of these government policies, two different Brigade approaches became evident. These two approaches are summarized below as ideal types rather than actual training models:

328

Original Brigade Approach	Modified Brigade Approach
1. Stresses rural development as primary goal.	1. Stresses manpower training for wage employment as primary goal.
2. Emphasizes production and employment. Training and work are inseparable and mutually supportive.	2. Emphasizes skill acquisition with production as secondary. Stresses the need to upgrade instructors and "professionalized" training.
3. Prefers "stay-put" training to create local employment.	3. Prefers "through-put" training to supplement labor force with skilled workers.
4. Values community development and self-reliance.	4. Values mobility and integration in the national education system.
5. Stresses cooperative production activities and participation of trainees in decision-making.	5. Stresses individual achievement, skill acquisition, and competence in government trade testing.
6. Emphasizes cost-recovery as an essential Brigade goal to assure financial autonomy and trainee "pride."	6. Sees cost-recovery operation as providing low-cost vocational training but also as detracting from needed skill-training time. Subsidies are used to reduce the necessity of production time in order to upgrade training quality.

    In choosing the overall technology for the project, the government selected the Modified Brigade Approach. The approach was estimated to be the least cost and most effective way to meet the project objectives when compared to the capital investment required by traditional vocational training centers and trade schools. A decision was made by the project planners to reinforce existing Brigades using the modified approach and increase enrollment through the following project components:

1. A National Brigade Development Center at the capital city--This would serve as the administrative and accounting center for the movement as a whole and as a training center to upgrade the skills and competence of instructors and administrators.

2. The establishment of three new Brigade Centers and the improvement of existing Centers--This would provide facilities for at least 1,000 new Brigade trainees and increase the total number of Brigade Centers to 15.*

The project was to be implemented through the National Coordinating Committee established in 1969 to act as liaison between the Ministry of Education and the Brigades, and approved by the local Trust. Location of the three Brigade Centers and the range of programs at any given location were to be decided by procedures maximizing community involvement. Trainees of the Brigade were to construct their own facilities, reducing the total cost of the project to U.S. $2.2 million.

Questions

1. What are some of the fundamental strengths of the Brigade approach described in this case as an alternative to traditional academic or vocational education?

2. Discuss the potential positive and negative impacts of government subsidies on brigade-type training.

3. Two "ideal types" of Brigade models were presented in the table above. Discuss each model's potential positive and negative impacts on the country's development. How can an effective training strategy be designed that would take into account both the rural

---

*At the time of the project design the number of Brigade Centers had reached twelve and the total number of Brigades was 66. There were 2,000 trainees in the Brigade system along with 200 managerial and instructional staff.

development goals and the national manpower needs for skilled labor?

4. In a proposed follow-up project, the country's government requested assistance for the creation of nine new agricultural Brigade units. If you were designing the training technology, would you select a through-put or a stay-put model? Why?

5. Discuss the role of values and ideology in the choice and design of technology for investment projects.

GLOSSARY

Active technology transfer: A mode of technology transfer where the transfer agent assists the potential user in its application.

Alternative technology: Hardware or software technology representing other than mainstream (modern) technology.

Appropriate technology: A generic term referring to technology that has the highest cost-benefit ratio, one that is most suitable to the technical, economic, social, cultural, and other conditions of the technology transfer recipient society or the receiving environment. It is one of the forms of intermediate (or alternative or progressive) technology and is characterized by such features as low investment per workplace, sparing use of natural resources, high potential for employment, organizational simplicity, but especially high adaptability to a particular sociocultural environment.

Appropriate technology movement: Those who strongly argue for sensitivity to sociocultural factors and the shunning of inappropriate technology in transfers.

Balance of payments: A summary of a country's international transactions over a given period of time, including commodity, service, technology, and other movements.

Brain drain: See Reverse technology transfer.

Capital drain: See Reverse technology transfer.

Communications: The process by which messages are transferred from a source to one or several receivers and which is at the heart of change, including technological change.

Community technology: A form of appropriate technology specifically tailored to the needs and capability of small urban or rural communities.

Contractual resource transfer: An arms-length sale or leasing of factors of production, e.g., technology.

Dependencia: The Spanish synonym of dependency which
relates to the vulnerability of the developing to
developed countries because of the former's economic and technological backwardness allegedly
flowing from their colonial experience. This perception of master-servant relationship between the
North and the South underlies the call for a New
International Economic Order.

Developed countries (DCs): Countries with relatively
high technology, per capita income, levels of
agricultural productivity, industrial production,
education, health, welfare, etc.

Developing countries (or less developed countries, or
LDCs): Countries which produce mainly raw materials or minerals for export but have inadequate
technological, economic, and/or social infrastructures. Despite many exceptions, generally their
per capita income, standard of living, and physical quality of life indices are low.

Development: The process through which a society becomes increasingly able to meet human needs and
assure a higher standard of living and physical
quality of life for its members. The type of
change that produces such higher levels of living
through better production methods and improved
technological, economic, and social organization.

Diffusion: The process by which new ideas are communicated to the members of a given social system.
Hence, the process by which innovations are communicated to the social system.

Direct investment: One of the most common mechanisms
of technology transfer usually involving the full
or partial ownership of a foreign subsidiary by a
parent firm.

East (as in East-West technology transfer): Broadly,
a nongeographic term referring to nonmarket, cen-
trally planned, communist countries. More nar-
rowly, communist East Europe and its economic
(Council of Mutual Economic Assistance, or COMECON,
or CMEA) and military (Warsaw Pact) alliances.

Engineering: The application of objective knowledge
to the creation of plans, designs, and means for

achieving desired objectives. It is the pervasive activity concerned with change in humanity's living environment and with the solution of problem situations which conflict with an individual's comfort, safety, or other challenges.

Feasibility study: (1) A study of applicability or desirability or any management or procedural system from the standpoint of advantages versus disadvantages in any given case. (2) A study to determine the time at which it would be practicable or desirable to install such a system when determined to be advantageous. (3) A study to determine whether a plan is capable of being accomplished successfully.

Gross Domestic Product (GDP): The aggregate of final, finished goods and services, excluding intermediate products, produced in a country in a year.

Gross National Product (GNP): The aggregate value of all goods and services produced by a country in a year comprising total expenditures by consumers, government, and gross private investment.

Group of 77 (G-77): A nongeographic term often equated with the "South," or developing countries. Originally, a group of less developed countries which issued a joint declaration at UNCTAD I (see below) in Geneva in 1964. The group now numbers over 120 countries and represents their caucus in the United Nations.

Hardware: A term borrowed from the computer industry to designate the physical embodiment of technology such as tools, appliances, implements, machines, devices, and equipment. In military technology, hardware designates weaponry for use in combat.

Horizontal technology transfer: A process that occurs when a useful idea moves laterally between different fields of application, often across organizational lines.

Inappropriate technology: Minimally, a technology having less than the highest cost-benefit ratio; by extension, a technology that is unsuitable to the economic, technical, social, cultural, or psychological conditions of the technology transfer recipient or the receiving environment.

335

Indirect investment: Another common form of technology transfer involving the transfer of knowledge through licenses, patents, etc., rather than direct investment in equipment or personnel.

Innovation: The conception of a new or improved product, process, or system followed by its/their commercialization. The application of a new product or process or technique with resulting commercial success.

Integrated technology transfer: The transfer of learning to do (technical know-how) together with learning to know (knowledge about knowledge), in contrast to the transfer of superficial, "rented," technology such as the mere change in the geographical location of a plant.

Interdependence: An increasingly obvious characteristic of the contemporary world where no society, however advanced or powerful, is totally immune from the consequences of actions and events happening elsewhere. Technology transfer tends to increase this state of interdependence.

Interface: Contact boundary between two solids, liquids, gases, or between two phases; by extension, interaction, interplay.

Intermediate technology: Technology that theoretically stands half-way between traditional and modern technology, all of these being relative terms. According to E. F. Schumacher, the originator of the concept, intermediate technology is labor-intensive, involves small-scale production, and decentralization.

International technology transfer or transnational technology transfer: The transfer of a new product or technique between different societies.

Intranational technology transfer: The transfer of a new product or technique within the same society.

Invention: The accidental or purposeful discovery of knowledge whose application may yield a useful result. The process of creating or developing a new product or system.

Know-how: The skill and capability necessary to convert technological knowledge into an economic reality. The ability to do something smoothly and efficiently.

Licensing: The grant of some but not necessarily all of the rights embraced in a copyright and thus a common mechanism of indirect technology transfer used as an alternative to direct investment.

Mainstream technology: See Modern technology.

Modern technology: Mainstream, or state-of-the-art, technology.

Multinational corporation (MNC): Also known as transnational enterprise (TNE), an MNC is a firm having production, research, or sales facilities in more than one country.

New International Economic Order (NIEO): The statement of development policies and objectives adopted at the Sixth Special Session of the United Nations General Assembly in 1974. The NIEO calls for equal participation of developing countries in international economic policy-making and the consequent massive transfer of resources, including technology, to them following such North-South dialogue for the sake of what they consider to be a more equitable distribution.

Newly industrialized countries (NICs): A group of developing countries which have witnessed sustained economic growth and technical advance.

North: A nongeographic term, synonymous with the "West," for the technologically and industrially advanced economies.

North-South dialogue: The discussion, often confrontational, between the industrial, Western market economies (North) and the developing economies (South) about trade preferences, economic and technical assistance, and terms of technology transfer. Initiated in 1974 with the Group of 77's call for a New International Economic Order.

Passive technology transfer: A mode of technology transfer where the transfer agent presents technology to the potential user but without assistance in its application--for instance, the transfer of knowledge as in the case of a patent or even a cookbook.

337

Patent: A government grant of a monopoly right that gives to the inventor or discoverer of a new and useful process, machine, manufacture, or composition of matter or a new and useful improvement thereof the exclusive right to make, use, or sell the invention or discovery for a specific period (17 years in the United States).

Private voluntary organizations (PVOs): Those which assist less privileged societies in various developmental matters, including technology transfer.

Research and development: The body of knowledge based on replicable scientific procedures concerning natural and social phenomena.

Reverse engineering: A process involving the plagiarism of technology by dismantling, examining, and emulating that of others, or through industrial espionage or other methods.

Reverse technology transfer: Technology transfer from developing to developed countries, especially in the form of trained personnel--the brain drain. Or, joint ventures abroad involving developing country participation in the form of capital goods, machinery, equipment--the capital drain.

Science: The knowledge that deals with human understanding of the real world, that is, with the inherent properties of space, matter, energy, and their interactions. It covers the empirical study of phenomena, the testing of hypotheses, and the formulation of generalizations that can be used as the basis of predictions.

Social change: An alteration in the structure and function of a social system. Social change involves three steps: Invention, diffusion, and structural reorganization.

Software: The nonmaterial dimensions of technology such as know-how, processes, programs, training, and other support services connected with hardware.

South: A nongeographic term applied to the economically and technologically less developed countries, often represented by the Group of 77.

338

Technical assistance: The transfer of industrial know-how to developing countries.

Technical feasibility: The state where available technical information, with perceived modifications, would provide a new technical capability.

Technological determinism: A belief in the existence of regular cause-and-effect relationships which control the world of technology.

Technological mastery: The ability to make effective use of technological knowledge.

Technologist: An individual, firm, or public agency which applies an invention in practice and demonstrates the feasibility of an idea.

Technology: Systematic knowledge and action, usually of industrial processes, but applicable to any recurrent activity closely related to science and engineering and viewed as providing the means of doing something desirable. Technology may be embodied in a physical reality (see Hardware) or in a method, technique, collection of techniques, or know-how, that is, the capacity to use technology (see Software).

Technology absorption: The assimilation of transferred technology by its recipients.

Technology adoption: The acceptance by a user of a technological practice common elsewhere or a different application of a given technique originally designed for another use. The latter is sometimes known as technology utilization, technology diffusion, or technology transfer.

Technology delivery system: The entire collection of private and public organizations involved in technology transfer.

Technology diffusion: The process by which an innovation spreads through a social or economic system. The multiple use of a product following research and development. Loosely, the term is at times used interchangeably with technology transfer.

Technology supplier: A source, private or public, national or international, which provides the technology to be transferred; the agent that tries to make technology transfer happen.

Technology transfer: The moving of technology, that is, of research and development results, into another setting at home or abroad for private or public user needs. It is the transplantation of technology from one set of well-defined conditions to another set in which at least one key variable may differ as may the way in which the recipient applies it. Or, the utilization of an existing technique in an instance where it has not previously been used (see also Technology diffusion).

Technology transfer barriers: The various obstacles—technical, economic, legal, social, cultural, psychological, or political—that may impede, restrict, or prevent smooth technology transfer.

Technology user: The consumer, private or public, national or international, actual or potential, who employs the technology transferred to satisfy a perceived or real need.

Technology utilization: The transfer of technology from one intended use to another, sometimes designated as technology diffusion or technology transfer.

Third World: Often used interchangeably with the terms developing countries, less developed countries, or the South. The Third World's main caucuses are the Group of 77, which acts as its economic voice, or the Movement of Nonaligned Nations, which acts as its political voice.

Traditional technology: Technology that is less than fully mechanized or automated or state-of-the-art.

Transnational: Transcending national borders and applicable to economic, social, political, and other systems.

Turnkey contract: An agreement in which an independent agent undertakes to furnish for a fixed price all materials and labor and to perform all functions needed to complete a project.

Turnkey technology: A ready-to-go technical installation.

Unbundling technology: The practice of separate pricing of software products and services from

340

equipment or other hardware items.

United Nations Conference on Trade and Development (UNCTAD): An agency of the United Nations system, first meeting in 1964 (UNCTAD I) in response to growing concerns among developing countries over their effort to bridge their economic and technological gap with the developed countries. Meetings were held again in 1968, 1972, 1976, 1979, and 1983 (UNCTAD II, III, IV, V, and VI) on North-South economic and technology-related issues.

Values: Preference for one state of reality over another. They specify what ought to be rather than what is by assigning a relative worth to objects and conditions.

Vertical technology transfer: A process that occurs when a new product or technique is communicated within an organization, usually from the research and development stage downstream to the manufacturing stage.

West: A nongeographic term, synonymous with the "North," referring to the economically, technologically, and industrially advanced countries of the Organization for Economic Cooperation and Development (OECD)--the 24 market economy countries of North America, Western Europe, Japan, and Australia. In a political or military context, the term refers to the Atlantic Alliance formed by NATO.

World Bank (or International Bank for Reconstruction and Development, or IBRD): A specialized agency in the United Nations system that makes long-term loans, mostly to develop infrastructural facilities, either directly to governments or with governments as guarantors.

342

BIBLIOGRAPHY

Agmon, Tamir and Charles P. Kindleberger (eds.), Multi-
    nationals from Small Countries. Cambridge, Mass.:
    The M.I.T. Press, 1977.

American Council of Voluntary Agencies for Foreign Ser-
    vice, Evaluation Sourcebook. New York: A.C.V.A.F.S.,
    1983.

Ayala, Hector, Transferencia de Tecnología de Companías
    Multinacionales a Países en Desarrollo: Casos en
    la Industria Colombiana. Bogota: CLADEA; and
    New York: Fund for Multinational Management Edu-
    cation, 1975 (in Spanish).

Baranson, Jack, North-South Technology Transfer. Mt.
    Airy, Md.: Lomond Publications, 1981.

_____, Technology and the Multinationals. Lexing-
    ton, Mass.: D. C. Heath and Co., 1978.

Baranson, Jack and Anne Harrington, Industrial Trans-
    fers of Technology by U.S. Firms under Licensing
    Arrangements: Policy, Practices and Conditioning
    Factors. Washington, D.C.: DEWIT, Inc., 1977.

Barbour, Ian G., Technology, Environment, and Human
    Values. New York: Praeger Publishers, 1980.

Baron, Christopher G. (ed.), Technology, Employment and
    Basic Needs in Food Processing in Developing Coun-
    tries. Oxford: Pergamon Press, 1980.

Behrman, Jack N. and Harvey W. Wallender, Transfers of
    Manufacturing Technology Within Multinational
    Enterprises. Cambridge, Mass.: Ballinger Pub-
    lishing Co., 1976.

Bereano, Philip L. (ed.), Technology as a Social and
    Political Phenomenon. New York: John Wiley and
    Sons, 1976.

Bertsch, Gary K., et al., "Decision Dynamics of Tech-
    nology Transfer to the U.S.S.R.," Technology in
    Society, Vol. 3, No. 4 (1981), pp. 412-415.

Bertsch, Gary K. and John R. McIntyre (eds.), National
    Security and Technology Transfer: The Strategic
    Dimensions of East-West Trade. Boulder, Colo.:

Westview Press, 1983.

Betz, Mathew, et al. (eds.), Appropriate Technology:
Choice and Development. Durham, N.C.: Duke University Press, 1984.

Bhalla, A.S. (ed.), Toward Global Action for Appropriate Technology. Oxford: Pergamon Press, 1979.

Bhattasali, B. N., Transfer of Technology Among the Developing Countries. Tokyo: Asian Productivity Organization, 1972.

Boretsky, M., "Export of U.S. Technology," Management of Science and Technology (1976), pp. 63-67.

Bourne, Malcolm C., "What Is Appropriate Intermediate Food Technology?" Food Technology, Vol. 32, No. 4 (April 1978), pp. 77-80.

Boyle, Godfrey, et al., The Politics of Technology. New York: Longman, 1977.

Boyle, Godfrey and Peter Harper (eds.), Radical Technology. London: Wildwood House, 1976.

Brasseur, Robert E., "Constraints in the Transfer of Knowledge," Focus, No. 3 (1976), pp. 12-19.

Braun, Ernest, et al., Assessment of Technology Decisions--Case Studies. London: Butterworths, 1979.

Carrick, R. J., East-West Technology Transfer in Perspective. Berkeley, Calif.: Institute of International Studies, University of California, 1978.

Casey, William J., "Technology Exchange with the U.S.S.R.: Current Status and Outlook," Research Management, Vol. 17, No. 4 (July 1974), pp. 7-9.

Chudson, Walter A., The International Transfer of Commercial Technology to Developing Countries. New York: UNITAR Research Report No. 13, 1974.

Clark, Norman, "The Multinational Corporation: The Transfer of Technology and Dependence," Development and Change, Vol. 6, No. 1 (1975), pp. 5-21.

Cole, Ralph I. and Sherman Gee (eds.), Proceedings of the Colloquium on Technology Transfer, 5-7 September 1973. Silver Spring, Md.: Publications

Division of the Naval Ordnance Laboratory, 1973.

Congdon, Robert J. (ed.), Introduction to Appropriate Technology. Emmaus, Pa.: Rodale Press, 1977.

Cooley, Mike, Architect or Bee? The Human/Technology Relationship. Boston, Mass.: South End Press, 1980.

Cooper, C., et al., "Choice of Techniques for Can-Making in Kenya, Tanzania and Thailand," in A. S. Bhalla (ed.), Technology and Employment in Industry. Geneva: International Labor Office, 1975, pp. 85-121.

Cotton, Frank E., Jr., "Some Interdisciplinary Problems in Transferring Technology and Management," Management International Review, Vol. 13, No. 1 (January 1973), pp. 59-65.

Dahlman, Carl L. and Larry E. Westphal, "The Meaning of Technological Mastery in Relation to Technology Transfer," The Annals of the American Academy of Political and Social Science, Vol. 458 (November 1981), pp. 12-26.

Danhof, Clarence H., Technology Transfer by People Transfer: A Case Study. Washington, D.C.: George Washington University, 1969.

Davidson, Harold F., et al. (eds.), Technology Transfer. Leiden, The Netherlands: Noordhoff, 1974.

Davis, Gregory H., Technology-Humanism or Nihilism: A Critical Analysis of the Philosophical Basis and Practice of Modern Technology. Washington, D.C.: University Press of America, 1981.

De Cubas, Jose, Technology Transfer and the Developing Nations. New York: Fund for Multinational Management Education, 1974.

Dickson, David, Alternative Technology and the Politics of Technical Change. London: Fontana/Collins, 1974.

Dorf, Richard C. and Yvonne L. Hunter (eds.), Appropriate Visions: Technology, the Environment and the Individual. San Francisco: Boyd and Fraser Publishing Co., 1978.

345

Driscoll, Robert E. and Harvey W. Wallender, III (eds.), Technology Transfer and Development: An Historical and Geographical Perspective. New York: Fund for Multinational Management Education, 1974.

Dunn, Peter D., Appropriate Technology: Technology with a Human Face. New York: Schocken Books, 1978.

Eckaus, Richard, Appropriate Technology for Developing Countries. Washington, D.C.: National Academy of Sciences, 1977.

Egea, Alejandro Nadel, "Multinational Corporations in the Operation and Ideology of International Transfer of Technology," Studies in Comparative International Development, Vol. 10, No. 1 (1975), pp. 11-29.

Ellul, Jacques, The Technological Society (translated from the French by John Wilkinson). New York: Random House, 1964.

Emmanuel, Arghiris, Appropriate or Underdeveloped Technology? (translated from the French by Timothy E. A. Benjamin). New York: John Wiley and Sons, 1982.

Evans, David K., "Applied Anthropological Methodology as a Contribution to Technology Transfer Programs within NATO," in Sherman Gee (ed.), Technology Transfer in Industrialized Countries. Alphen aan den Rijn, The Netherlands: Sijthoff and Noordhoff International Publishers, 1979, pp. 343-359.

Evans, Donald D. and Laurie Nogg Adler (eds.), Appropriate Technology for Development: A Discussion and Case Histories. Boulder, Colo.: Westview Press, 1979.

Fleron, Frederic J., Jr. (ed.), Technology and Communist Culture: The Socio-Cultural Impact of Technology under Socialism. New York: Praeger Publishers, 1977.

Frame, J. Davidson, International Business and Global Technology. Lexington, Mass.: D. C. Heath and Co., 1983.

Galtung, Johan, Development, Environment and Technology: Towards a Technology of Self-Reliance.

Geneva: United Nations Conference on Trade and Development, document TD/B/C.6/23/Rev. 1, 1979.

Gartner, Joseph and Charles S. Naiman, "Overcoming the Barriers to Technology Transfer," Research Management, Vol. 19, No. 3 (March 1976), pp. 22-28.

Gee, Sherman, "Military-Civilian Technology Transfer: Progress and Prospects," Defense Management Journal, Vol. 11, No. 2 (April 1975), pp. 46-51.

_____ (ed.), Technology Transfer in Industrialized Countries. Alphen aan den Rijn, The Netherlands: Sijthoff and Noordhoff International Publishers, 1979.

_____, Technology Transfer, Innovation, and International Competitiveness. New York: John Wiley and Sons, 1981.

Germidis, Dmitri, Transfer of Technology by Multinational Corporations. Paris: Organization for Economic Cooperation and Development, 1977.

Ghatak, Subrata, et al., Technology Transfer to Developing Countries: The Case of the Fertilizer Industry. Greenwich, Conn.: JAI Press, Inc., 1981.

Ginzberg, E., Technology and Social Change. New York: Columbia University Press, 1964.

Girling, Robert, "Mechanisms of Imperialism: Technology and the Dependent State," Latin American Perspectives, Vol. 3, No. 4 (1976), pp. 54-64.

Goulet, Denis, "The Paradox of Technology Transfer," The Bulletin of the Atomic Scientists, Vol. 31, No. 6 (June 1975), pp. 39-46.

_____, "The Suppliers and Purchasers of Technology: A Conflict of Interests," International Development Review, Vol. 18 (March 3, 1976), pp. 14-20.

_____, The Uncertain Promise: Value Conflicts in Technology Transfer. New York: IDOC/North America, 1977.

Granger, John V., Technology and International Relations. San Francisco: W. H. Freeman and Co.,1979.

347

Greiner, Ted, The Promotion of Bottle Feeding by Multi-
national Corporations: How Advertising and the
Health Professions Have Contributed. Ithaca, N.Y.:
Cornell University, Program on International Nutri-
tion and Development Policy, Cornell International
Nutrition Monograph Series No. 2, 1975.

Gruber, William H. and Donald G. Marquis (eds.), Factors in the Transfer of Technology. Cambridge,
Mass.: The M.I.T. Press, 1969.

Haden-Guest, Anthony, Down the Programmed Rabbit Hole:
Travels Through Muzak, Hilton, Coca-Cola, Walt
Disney and Other World Empires. London: Hart-
Davis, MacGibbon, Ltd., 1972.

Hanson, Philip, Trade and Technology in Soviet-Western
Relations. New York: Columbia University Press,
1981.

Harrison, Paul, "Small Is Appropriate," New Scientist,
Vol. 82, No. 1149 (April 5, 1979), pp. 14-16.

Hawthorne, Edward P., The Transfer of Technology.
Paris: Organization for Economic Cooperation and
Development, 1971.

Hayden, Eric W., Technology Transfer to Eastern Europe:
U.S. Corporate Experience. New York: Praeger
Publishers, 1976.

Heston, Alan W. and Howard Pack (eds.), "Technology
Transfer: New Issues, New Analysis," The Annals
of the American Academy of Political and Social
Science, Vol. 458 (November 1981).

Higgins, James A., "Technology Transfer: A Key to
Productivity," Defense Systems Management Review,
Vol. 2, No. 1 (Winter 1979), pp. 7-9.

Hodges, Wayne L. and Matthew A. Kelly (eds.), Technological Change and Human Development. Ithaca,
N.Y.: Cornell University Press, 1970.

Hoelscher, Harold E., Technology and Man's Changing
World: Some Thoughts on Understanding the Interaction of Technology and Society. Beirut, Leba-
non: American University of Beirut, 1980.

Hollick, Malcolm, "The Appropriate Technology Movement
and Its Literature: A Retrospect," Technology in

Society, Vol. 4, No. 3 (1982), pp. 213-229.

Holliday, George D., Technology Transfer to the U.S.S.R.
1928-1937 and 1966-1975: The Role of Western Technology in Soviet Economic Development. Boulder,
Colo.: Westview Press, 1979.

Horvath, Janos, Chinese Technology Transfer to the
Third World. New York: Praeger Publishers, 1976.

Hough, Granville W., Technology Diffusion. Mt. Airy,
Md.: Lomond Publications, 1975.

Illich, Ivan D., Energy and Equity. New York: Harper
and Row, 1974.

_____, The Limits to Medicine: Medical Nemesis.
New York: Penguin, 1978.

Independent Commission on International Development
Issues, North-South: A Program for Survival
(Brandt Report). Cambridge, Mass.: The M.I.T.
Press, 1980.

Jequier, Nicolas (ed.), Appropriate Technology: Problems and Promises. Paris: Organization for Economic Cooperation and Development, 1976.

_____, Technology Transfer and Appropriate Techno-
logy. Singapore: European Institute of Business
Administration, Seminar on the Management and
Transfer of Technology, 1978.

Kamenetzky, Mario, "Choice and Design of Technologies
for Investment Projects." Washington, D.C.: The
World Bank, April 4, 1982 (mimeo).

Katz, Jorge M., Importación de Tecnología, Aprendizaje
Local e Industrialización Dependiente. Buenos
Aires, Argentina: Instituto Torcauto de Tella,
1972 (in Spanish).

Keddie, James and William Cleghorn, Brewing in Developing Countries. Edinburgh, Scotland: Scottish
Academic Press Ltd., 1979.

Keller, Robert T., "A Look at the Socio-Technological
System," California Management Review, Vol. 15
(Fall 1972), pp. 86-91.

Koeppe, D. F., Technology Transfer: Theory and

Applications. Dallas, Tex.: Twenty-First Century Corp., 1977.

Konz, Leo E., The International Transfer of Commercial Technology: The Role of the Multinational Corporation. New York: Arno Press, 1980.

Kumar, Krishna (ed.), Bonds Without Bondage: Explorations in Transcultural Interactions. Honolulu: University Press of Hawaii, 1979.

_____, The Social and Cultural Impacts of Transnational Enterprises. Sydney, Australia: The University of Sydney, Transnational Corporations Research Project No. 6, 1979.

_____ (ed.), Transnational Enterprises: Their Impact on Third World Societies and Cultures. Boulder, Colo.: Westview Press, 1980.

Kumar, Krishna and Maxwell G. McLeod (eds.), Multinationals from Developing Countries. Lexington, Mass.: D. C. Heath and Co., 1981.

Laarman, J., et al., Choice of Technology in Forestry: A Philippine Case Study. Quezon City, Philippines: New Day Publishers, 1981.

Lall, Sanjaya, "Developing Countries as Exporters of Industrial Technology: A Preliminary Analysis," in Herbert Giersch (ed.), International Economic Development and Resource Transfer Workshop 1978. Tubingen, West Germany: J. C. B. Mohr, 1979, pp. 589-610.

Lewis, W. H., "Arms Transfers and the Third World," in U.S. Arms Control and Disarmament Agency, World Military Expenditures and Arms Transfers 1969-1978. Washington, D.C.: U.S. Government Printing Office, 1982.

Liebrenz, Marilyn L., Transfer of Technology: U.S. Multinationals and Eastern Europe. New York: Praeger Publishers, 1982.

Lipscombe, Joan and Bill Williams, Are Science and Technology Neutral? London: Butterworths, 1979.

Long, Frank, "The Management of Technology Transfer to Public Enterprises in the Caribbean," Technology in Society, Vol. 5, No. 1 (1983), pp. 69-82.

Long, Franklin and Alexandra Oleson (eds.), Appropriate
    Technology and Social Values: A Critical Apprais-
    al. Cambridge, Mass.: Ballinger Publishing Co.,
    1980.

Lucas, Barbara A. and Stephen Freedman (eds.), Techno-
    logy Choice and Change in Developing Countries:
    Internal and External Constraints. Dublin, Ire-
    land: Tycooly International Publishing Ltd., 1983.

McRobie, George, Small Is Possible. New York: Harper
    and Row, 1981.

Manning, G. K., Technology Transfer: Successes and
    Failures. San Francisco: San Francisco Press,
    1974.

Manser, W. A. P. and Simon Webley, Technology Transfer
    to Developing Countries. London: The Royal In-
    stitute of International Affairs, 1979.

Mansfield, Edwin, Technological Change. New York:
    W. W. Norton and Co., 1971.

Mascarenhas, R. C., Technology Transfer and Develop-
    ment. Boulder, Colo.: Westview Press, 1982.

Mason, R. Hal, "The Selection of Technology: A Con-
    tinuing Dilemma," Columbia Journal of World Busi-
    ness, Vol. 9 (Summer 1974), pp. 29-34.

Mathieson, Raymond S., Japan's Role in Soviet Economic
    Growth: Transfer of Technology Since 1965. New
    York: Praeger Publishers, 1979.

Mattelart, Armand, "Cultural Imperialism in the Multi-
    nationals Age," Instant Research on Peace and Vio-
    lence, Vol. 6, No. 4 (1976), pp. 160-174.

Mitchel, R. J. (ed.), Experiences in Appropriate Tech-
    nology. Ottawa: The Canadian Hunger Foundation,
    1980.

Mogavero, Louis N. and Robert S. Shane, What Every
    Engineer Should Know About Technology Transfer
    and Innovation. New York: Marcel Dekker, Inc.,
    1982.

Muller, Herbert J., The Children of Frankenstein: A
    Primer on Modern Technology and Human Values.
    Bloomington: Indiana University Press, 1970.

Muller, M., The Baby Killer. London: War on Want
Pamphlet, 1974.

Murphy, Kathleen J., "Third World Macroprojects in the
1970s: Human Realities--Managerial Responses,"
Technology in Society, Vol. 4, No. 2 (1982), pp.
131-144.

El-Namaki, M., et al., Mission Report on Appropriate
Technology in Tanzania. Delft, The Netherlands:
Center for Appropriate Technology, 1979.

Nasbeth, L. and G. F. Ray, The Diffusion of New Industrial Processes. Aberdeen, Mass.: Cambridge University Press, 1974.

National Aeronautics and Space Administration, Spinoff
1981. Washington, D.C.: U.S. Government Printing
Office, 1982.

Nau, Henry R., Public Policy and Technology Transfer.
New York: Fund for Multinational Management Education, 1978.

Nevin, Paul, The Survival Economy: Micro-Enterprises
in Latin America. Cambridge, Mass.: Accion In-
ternational/AITEC, 1982.

Norman, Colin, The God That Limps: Science and Technology in the Eighties. New York: W. W. Norton
and Co., 1981.

Norman, Henry R. and Patricia Blair, "The Coming Growth
in 'Appropriate' Technology," Harvard Business Review, Vol. 60, No. 6 (November/December 1982),
pp. 62-63 and 66.

Ohiorhenuan, John F. E., Social, Environmental and Eco-
nomic Aspects of Technology Transfer in Jamaican
Tourism. Geneva: United Nations Conference on
Trade and Development, UNEP/UNCTAD document TD/
B.C6/49, November 28, 1979.

Organization for Economic Cooperation and Development,
Biotechnology: International Trends and Perspectives. Paris: OECD, 1982.

_____, Choice and Adaptation of Technology in De-
veloping Countries. Paris: OECD Development
Centre, 1974.

_____, North/South Technology Transfer: The Adjustments Ahead. Paris: OECD, 1981.

Ozawa, Terutomo, Transfer of Technology from Japan to Developing Countries. New York: United Nations Institute for Training and Research, 1971.

Parker, Robert N., "Technology Exchange with the U.S.S.R.: National Security Issues," Research Management, Vol. 17, No. 4 (July 1974), pp. 12-13.

Pavitt, Keith, "The Multinational Enterprise and the Transfer of Technology," in John H. Dunning (ed.), The Multinational Enterprise. London: Allen and Unwin, 1971.

Peters, E. Bruce, "Cultural and Language Obstacles to Information Transfer in the Scientific and Technical Field," Management International Review, Vol. 15, No. 1 (January 1975), pp. 75-88.

Pickett, James, et al., "The Choice of Technology, Economic Efficiency and Employment in Developing Countries," World Development, Vol. 2, No. 3 (March 1974), pp. 47-54.

Piekarz, Rolf R. (ed.), The Effects of International Technology Transfers on the U.S. Economy. Washington, D.C.: National Science Foundation, 1973.

Poats, Rutherford M., Technology for Developing Nations. Washington, D.C.: The Brookings Institution, 1972.

Prasad, A. J., "Export of Technology from India." Unpublished Ph.D. dissertation, Columbia University, New York, 1978.

Proposal for a Program in Appropriate Technology: Report of the Agency for International Development to the United States Congress Pursuant to Section 197 of the Foreign Assistance Act. Washington, D.C.: U.S. Government Printing Office, 1976.

Public Policy and Technology Transfer. New York: Fund for Multinational Management Education, 1978, vols. 1-4.

Ramaswamy, G. S., "The Transfer of Technology Among Developing Countries," Focus, No. 2 (1976), pp. 7-10.

Ramesh, Jairam and Charles Weiss (eds.), Mobilizing Technology for World Development. New York: Praeger Publishers, 1979.

Ranis, Gustav, "Appropriate Technology: Obstacles and Opportunities," in Samuel M. Rosenblatt (ed.), Technology and Economic Development: A Realistic Perspective. Boulder, Colo.: Westview Press, 1979.

Reddy, A. K. N., Technology, Development and the Environment: A Reappraisal. Nairobi: United Nations Environment Programme, 1979.

Rhee, Yung W. and Larry E. Westphal, "A Micro, Econometric Investigation of Choice of Technology," Journal of Development Economics, Vol. 4, No. 3 (September 1977), pp. 205-237.

Richardson, Jacques (ed.), Integrated Technology Transfer. Mt. Airy, Md.: Lomond Publications, 1979.

Robinson, Austin (ed.), Appropriate Technologies for Third World Development: Proceedings of a Conference Held by the International Economic Association in Teheran, Iran. New York: St. Martin's Press, 1979.

Rogers, Everett M., Diffusion of Innovations. New York: The Free Press, 1968.

Rogers, Everett M. and F. F. Shoemaker, Communication of Innovations: A Cross-Cultural Approach. New York: The Free Press, 1971.

Rokeach, Milton, The Nature of Human Values. New York: The Free Press, 1973.

Root, William A., "Controls on West-to-East Technology Transfer," Defense Systems Management Review, Vol. 2, No. 1 (Winter 1979), pp. 44-52.

Rosenblatt, Samuel M. (ed.), Technology and Economic Development: A Realistic Perspective. Boulder, Colo.: Westview Press, 1979.

Rosenbloom, Richard S. and Francis W. Wolek, Technology and Information Transfer. Boston, Mass.: Harvard University Press, 1970.

Rybczinski, Witold, Paper Heroes: A Review of

Appropriate Technology. Garden City, N.Y.: Anchor
Press/Doubleday, 1980.

Sagafi-nejad, Tagi, _Transfer of Technology from Egypt_.
Vienna: United Nations Industrial Development
Organization, document UNIDO/IS.362, 1982.

_____, et al., _Controlling International Techno-
logy Transfer_. New York: Pergamon Press, 1981.

Sagafi-nejad, Tagi and Robert Belfield, _Transnational
Corporations, Technology Transfer, and Develop-
ment: A Bibliographic Sourcebook_. New York:
Pergamon Press, 1980.

Saint-Rossy, D. T., et al., _International Transfer of
Technology--An Evaluation of the Education and
Training Component_. Evanston, Ill.: The Technological Institute of Northwestern University, May
1976.

Salisbury, R. F., _From Stone to Steel: Economic Consequences of a Technological Change in New Guinea_.
Melbourne, Australia: Melbourne University Press,
1962.

Schumacher, E. F., "The Case for Intermediate Technology," in G. M. Meier (ed.), _Leading Issues in
Economic Development_ (2nd ed.). New York: Oxford University Press, 1970, pp. 355-359.

_____, _Good Work_. New York: Harper and Row,
1979.

_____, _Small Is Beautiful: Economics as if People
Mattered_. New York: Harper and Row, 1973.

Seitz, Frederick, "The Role of Universities in the
Transnational Interchange of Science and Techno-
logy for Development," _Technology in Society_,
Vol. 4, No. 1 (1982), pp. 33-40.

Science, Technology, and Human Values. Cambridge,
Mass.

Seurat, Silvere, _Technology Transfer: A Realistic
Approach_. Houston, Tex.: Gulf Publishing Co.,
1979.

Singer, Hans W., "Appropriate Technology for a Basic
Human Needs Strategy," _International Development_
355

Review, Vol. 19, No. 2 (1977), pp. 8-11.

_____, "The Transfer of Technology in LDCs," _Inter-Economics_, No. 1 (January 1974), pp. 14-17.

Smith, Charles H. III, _Japanese Technology Transfer to Brazil_. Ann Arbor, Mich.: University Microfilms International, 1981.

Sobeslavsky, V. and P. Beazley, _The Transfer of Technology to Socialist Countries: The Case of the Soviet Chemical Industry_. Westmead, England: Gower Publishing Co., 1980.

Solo, Robert A., _Organizing Science for Technology Transfer in Economic Development_. East Lansing: Michigan State University Press, 1975.

Spencer, Daniel L., _Technology Gap in Perspective: Strategy of International Technology Transfer_. New York: Spartan Books, 1970.

Stavrianos, L. S., _The Promise of the Coming Age_. San Francisco, Calif.: W. H. Freeman, 1976.

Steele, Lowell W., "Barriers to International Technology Transfer," _Research Management_, Vol. 17, No. 1 (January 1974), pp. 17-21.

Stewart, Frances, _Macro-Policies for Appropriate Technology: An Introductory Classification_. Washington, D.C.: Appropriate Technology International, 1983.

_____, _Technology and Underdevelopment_ (2nd ed.). London: The Macmillan Press, 1978.

Stover, William J., _Information Technology in the Third World: Can I.T. Lead to Humane National Development?_ Boulder, Colo.: Westview Press, 1984.

Streeten, Paul, "The Multinational Corporation and the Nation State," in Paul Streeten (ed.), _The Frontiers of Development Studies_. New York: John Wiley and Sons, 1972.

_____, "Self-Reliant Industrialization," in Charles K. Wilber (ed.), _The Political Economy of Development and Underdevelopment_ (2nd ed.). New York: Random House, 1979, pp. 281-296.

Sutter, Rolf, "Technology Transfer into LDCs," Inter-Economics, No. 12 (December 1974), pp. 380-384.

Sutton, A. C., Western Technology and Soviet Economic Development, 1917-1965. Stanford, Calif.: Hoover Institution, Vols. 1-3, 1968-73.

Szyliowicz, Joseph S. (ed.), Technology and International Affairs. New York: Praeger Publishers, 1981.

Technology Management Action. Palo Verdes, Pa.

Technology Transfer. Saskatoon, Sask., Canada.

Teece, David J., The Multinational Corporation and the Resource Cost of International Technology Transfer. Cambridge, Mass.: Ballinger Publishing Co., 1976.

Tilton, J. E., International Diffusion of Technology: The Case of Semiconductors. Washington, D.C.: The Brookings Institution, 1971.

Timmer, C. Peter, et al., The Choice of Technology in Developing Countries: Some Cautionary Tales. Cambridge, Mass.: Harvard University Press, Harvard Studies in International Affairs No. 12, 1975.

Todd, John and Nancy, Tomorrow Is Our Permanent Address. New York: Harper and Row, 1979.

Todd, Nancy and Jack (eds.), The Book of the New Alchemists. New York: E. P. Dutton, 1977.

TRANET Newsletter. Rangeley, Me.

Turner, Louis, "The International Division of Leisure, Tourism, and the Third World," World Development, Vol. 4, No. 3 (1976), pp. 253-260.

Twiss, B. C., Managing Technological Innovation (2nd ed.). New York: Longman, 1980.

United Nations Association of the United States of America, Issues Before the 37th General Assembly of the United Nations. New York: UNA-USA, 1982, pp. 90-95.

United Nations Conference on Trade and Development, Major Issues Arising from the Transfer of

Technology to Developing Countries. New York: United Nations, 1975.

_____, The Reverse Transfer of Technology: Economic Effects of the Outflow of Trained Personnel from Developing Countries. Geneva: UNCTAD, document TD/B/AC.11/25, Rev. 1, 1975.

_____, "Select Bibliography of Documents on Transfer and Development of Technology." Geneva: UNCTAD, document TD/B/C.6 INF.2, Rev. 2, November 7, 1980 (mimeo).

_____, Transfer and Development of Technology in Egypt. Geneva: UNCTAD, document UNCTAD/TT/AS/7, 1980.

_____, Transfer of Technology: Its Implications for Development and Environment. Geneva: UNCTAD, document TD/B/C.6/22, 1978.

United Nations Department of Economic and Social Affairs, The Acquisition of Technology from Multinational Corporations by Developing Countries. New York: United Nations, 1974.

United Nations Industrial Development Organization, Conceptual and Policy Framework for Appropriate Industrial Technology. Vienna: UNIDO, Monographs on Appropriate Industrial Technology No. 1, 1979.

_____, Technologies from Developing Countries. Vienna: UNIDO, Development and Transfer of Technology Series No. 7, 1978.

United States Congress, Joint Economic Committee, Issues in East-West Commercial Relations: A Compendium of Papers. Washington, D.C.: U.S. Government Printing Office, 1979.

Utterback, J. M., "Innovation in Industry and the Diffusion of Technology," Science, No. 183 (February 15, 1974), pp. 620-626.

Vacca, Robert, Modest Technologies for a Complicated World. Oxford: Pergamon Press, 1980.

Villamil, J. J. (ed.), Transnational Capitalism and National Development. Atlantic Highlands, N.J.: Humanities Press, 1977.

Vogeler, Ingolf and Anthony R. de Souza (eds.), Dialectics of Third World Development. Montclair, N.J.: Allanheld, Osmun, 1980.

Voll, Sarah P., A Plough in Field Arable: Western Agribusiness in the Third World. Hanover, N.H.: University Press of New England, 1980.

Wallender, Harvey W. III, Technology Transfer and Management in the Developing Countries: Company Cases and Policy Analyses in Brazil, Kenya, Korea, Peru, and Tanzania. Cambridge, Mass.: Ballinger Publishing Co., 1979.

Weiss, Charles, Jr., Mobilizing Technology for Developing Countries. Washington, D.C.: The World Bank, World Bank Reprint Series No. 95, 1979.

Wells, Alan F., Picture-Tube Imperialism? The Impact of U.S. Television on Latin America. Maryknoll, N.Y.: Orbis Books, 1972.

Wells, Louis T., Jr., "Economic Man and Engineering Man," Development Digest, Vol. 14 (October 1976), pp. 71-83.

Weinstein, Franklin, "Multinational Corporations and the Third World: The Case of Japan and Southeast Asia," International Organization, Vol. 30, No. 3 (Summer 1976), pp. 373-404.

Wionczek, Miguel, "Notes on Technology Transfer Through Multinational Enterprises in Latin America," Development and Change, Vol. 7, No. 2 (1976), pp. 135-155.

Zahlan, Antoine B. and Rosemarie Said Jahlan (eds.), Technology Transfer and Change in the Arab World. New York: Pergamon Press, 1978.

Zaleski, Eugene and Helgard Wienert, Technology Transfer Between East and West. Paris: Organization for Economic Cooperation and Development, 1980.

362

365